SCIENCE LITERACY IN PRIMARY SCHOOLS AND PRE-SCHOOLS

CLASSICS IN SCIENCE EDUCATION

Volume 1

Series Editor:

Karen C. Cohen

SCIENCE LITERACY IN PRIMARY SCHOOLS AND PRE-SCHOOLS

By

Haim Eshach

Ben Gurion University of the Negev, Beer Sheva, Israel

A.C.I.P. Catalogue record for this book is available from the Library of Congress.

ISBN–10 1–4020–4641–3 (HB)
ISBN–13 978–1–4020–4641–4 (HB)
ISBN–10 1–4020–4674–X (e-book)
ISBN–13 978–1–4020–4674–2 (e-book)

Published by Springer,
P.O. Box 17, 3300 AA Dordrecht, The Netherlands.

www.springer.com

Printed on acid-free paper

Printed in the Netherlands

This book is dedicated to my three children, Omry, Shaked, and Ohad. They were the reason that I started taking an interest in early childhood science education 12 years ago. At first, I just wanted to do some science activities with them, but then I was swept into the early childhood science education concept. In this regard one may say that, like Saul, who went out looking for horses but found an entire kingdom, I also found a whole world in early science education. I hope that as Omry, Shaked, and Ohad have inspired me in writing this book, this book will be the source of a science education, inspiring them with the desire, enthusiasm, and eagerness to know and to learn.

I also want to thank my wife, Orly, whose love never failed to give me strength . . .

CONTENTS

PREFACE

Science is more than a compilation of facts and figures, although one would not know that from observing classroom lessons in science in elementary schools in many parts of the world. In fact, there are those who argue that science is not appropriate subject content for the early grades of elementary school. There are many schools in which science is simply not present in the earliest grades. Even where science is taught in the earliest grades, it is often a caricature of science that is presented to the children.

This book offers a vigorous, reasoned argument against the perspective that science doesn't belong in the early grades. It goes beyond that in offering a view of science that is both appropriate to the early grades and faithful to the nature of the scientific enterprise. Dr. Eshach is not a voice in the chorus that claims young children's developmental lack of readiness for such study.

He believes, as do I, that in order to learn science one must do science. At the heart of the doing of science is the act of exploration and theory formation. To do science, we must explore the ways in which the world around us looks, sounds, smells, feels, and behaves.

But science is more than a catalog of data. Doing science involves us in the search for causal, rational explanation. Children, from the earliest grades, must come to value the importance not only of collecting evidence, but reasoning from that evidence to generate testable hypotheses and predictions.

To become responsible citizens in an increasingly science- and technology-driven world, young people must develop a range of "habits of mind" that will lead them to understand the power and the peril of idealization and the importance of analogical thinking. They must internalize the need for intellectual humility peppered with a healthy dose of skepticism. They must learn to look for causes and anticipate possible ranges of results.

This is a tall order. Starting down this road can't start early enough. We are fortunate that to help us we have an astute guide in this book and its author.

Judah L. Schwartz
Visiting Professor of Education and Research Professor of Physics & Astronomy, Tufts University
Emeritus Professor of Engineering Science & Education, MIT
Emeritus Professor of Education, Harvard University

INTRODUCTION

On December 26, 2004, Tilly Smith, a 10-year-old girl from Oxshott, Surrey in England was on holiday with her family in Phuket beaches in Thailand. On that day, a tsunami hit Phuket's beaches. More than 200,000 people were killed, making it one of the deadliest disasters in modern history. Today we all have a notion of what tsunami is — a series of huge waves following an undersea disturbance, such as an earthquake or volcano eruption. The term *tsunami* comes from the Japanese language meaning *harbor* ("tsu", 津) and *wave* ("nami", 波 or 浪) (from Wikipedia, the free encyclopedia, http://en.wikipedia.org/wiki/Tsunami). The tsunami that hit Phuket's beaches originated from the 2004 Indian Ocean earthquake, known by the scientific community as the Sumatra-Andaman undersea earthquake.

By chance, the little girl from England, Tilly Smith, had just learned about the tsunami phenomenon in school two weeks prior to her vacation. Having learned this lesson in a scientifically oriented part of a geography class, she recognized that the "funny" behavior of the water, the bubbles and the tide which went out all of a sudden might portend a tsunami. She told her parents who fortunately listened and urged others off Maikhao Beach and to high ground. The beach was cleared and about 100 lives were saved. The science lesson saved lives! This should be a lesson to everyone as to how science ideas can be understood by young children, and moreover, that science education as early as primary school might impact children's behavior in real life situations. Tilly's story is fascinating; however, we do not need catastrophe heroes to realize how science and technology help children to deal with problems which they are confronted with daily. As a parent and as an educator who now has more than 12 years of experience in teaching science to preschool teachers, I have reached the conclusion that children living in the 21st century not only deserve but also need to receive science education as early as kindergarten.

Unfortunately, the current state of science education in primary schools continues to be a cause of concern (Harlen, 1997). Based on the U.S. students' performance on the Third International Mathematics and Science Study (TIMSS), Schmidt *et al.* (1997), for instance, state that there is "no single coherent vision of how to educate today's children in . . . science" (p. 1). Harlen (1997), reviewing the literature reaches a similar conclusion according to which, "many children are experiencing narrow and impoverished learning opportunities which hardly qualify to be described as science education" (p. 335). In addition, Howes (2002) argues that a prototypical picture persists of pre-service elementary teachers as lacking what it takes to teach science.

This book aims at changing this situation. It urges countries, wherever they are, to invest energy, thought and attention to early science education. As I said before, that is not the present case; indeed, in most places there is a lamentable lack of commitment to the idea of early science education. In Israel, for instance, there is a framework syllabus for science in kindergarten; however, it is not a compulsory one.

Teachers can choose whether or not they want to include science in their kindergarten's activities. Recognizing the need to begin science education in the early stages of life, the Israeli ministry of education appointed me to be the chairperson of a national committee which will decide what will be the core compulsory curriculum in science and technology education in preschool. In 2007 the committee is expected to provide the ministry with the final document. Every child in Israel will then start his or her scientific enterprise already at kindergarten! I am really eager and excited to see the change. I find this a crucial and important step which will enable us to provide the children of today with the knowledge and skills necessary for them, the adults of tomorrow, to deal with natural catastrophes like the real tsunami in Thailand. Moreover, a good science education — by which I mean one that will nurture scientific thinking skills and inculcate in children the desire and passion to know and learn — will provide us with hope that our children, our next generation, will create a better world for people to live in. I am however, well aware that *bad* science education can sometimes be worse than no science education at all. Thus, this book, while being an unreserved call for early science education, is also a call for sober caution as we pursue it: we want to assure that the path we follow for this adventure is safe.

GUIDE TO THE BOOK

The book begins with a more philosophical question, which is also the name of chapter 1, *Should Science be Taught in Early Childhood?* To start a good discussion on questions such as to how science should be taught, one should first be deeply convinced as to the importance of science education already at first years of life. The chapter shows why the typical reasons given by educators are problematic. In addition, six justifications are provided for exposing young children to science, which at least make taking up the enterprise more reasonable than rejecting it. From here the book moves into more practical aspects, and deals with ways in which one can teach science to K-2 children. Chapter 2, *How Should Science be Taught in Early Childhood?* presents and discusses the following approaches: inquiry-based teaching; learning through authentic problems; preference of the *psychological* rather than the *logical* order; scaffolding; situated learning; learning through projects; and non-verbal knowledge. Chapter 3, *When Learning Science by Doing meets Design and Technology*, continues the more practical side of the book. Although also rich with theory, it presents and discusses a fresh and novel approach to how technology, especially designing, building, evaluating and redesigning simple artifacts, may be an efficient vessel for promoting science learning. Chapter 4, *From the Known to the Complex: The Inquiry Events Method as a Tool for K-2 Science Teaching*, presents the following interesting idea: K-2 scientific curricula should put the needs of the teachers in the center and not only those of children, as is usually the case. One such approach, which is presented in the chapter, is called the "Inquiry Event." The chapter is divided into three parts. The first part introduces the method and the rational. The second is a research on K-2 educators' efficacy belief change regarding science teaching as a result of participating in a workshop dealing with the method. The third part of this chapter continues the inquiry on the IE approach and evaluates the IE teaching method in two Israeli's kindergartens. The question of whether and how early science education should be the business of K-2 is a question of what should be in, and, by implication, out of school. Therefore, the final chapter, Chapter 5, *Bridging In-School and Out-of-School Learning: Formal, Non-Formal, and In-formal* tries to draw more precisely the boundaries between in-school and out-of-school learning. By doing so, I hope to bring a more comprehensive view of K-2 science education. The chapter explains the notion of out-of-school learning and provides a thorough review on its characteristics. In addition, it suggests some interesting ways to bridge in and out-of school learning.

Although this book provides ample examples on how to bring the theories presented in the different chapters into practice, it is still meant to be a more theoretical exposition, and not a comprehensive guide to be used as a curriculum for K-2 science teaching. This book should serve well for researchers and those who develop and design K-2 science teaching materials and curricula, along with K-2 school teachers. Also, parents who are interested in science education might find it to be an inspirational source to helping them get involved with their own children's science education.

ACKNOWLEDGMENTS

This book was written during the academic years 2003–2005 when I was generously supported by the Sacta Rashi foundation as a Guastela fellow. I would not have had the stability and security without this support, for which I am grateful. I would like to thank Dr. Michael N. Fried, my colleague, who is also a good friend. Writing the first chapter of the book with Michael served as a good starting point which inspired me throughout the whole writing process. In addition, as a historian of mathematics as well as an education researcher, Michael always brought interesting and insightful issues to the many conversations we had on educational issues, which enriched my perspectives. He had many helpful suggestions which I always found to be of great help. I also want to thank Mr. Roy Golombick for his editorial suggestions. Roy is a scientist himself, and therefore was an excellent choice for editing this book. His critiques were valuable contributions. I would also like to thank Ms. Liat Bloch whom I guided in her M.Sc. thesis. Her dedication to her research, as well as her passion for early childhood science education enabled me to evaluate the Inquiry Events method which is presented in chapter 4. The third part of this chapter was written with Liat. The department of Science Education and Technology at Ben Gurion University, its present Director, Prof. Miriam Amit, as well as her predecessor, Prof. Shlomo Vinner also merit my gratitude for their encouragement and support. Finally, but not least, I would like to thank Professor Karen C. Cohen for her continual help, advice and inspiration, without which this project would never have been possible.

CHAPTER 1

SHOULD SCIENCE BE TAUGHT IN EARLY CHILDHOOD?[1]

Early in his life, the physicist Enrico Fermi resolved "to spend at least one hour a day thinking in a speculative way" (Ulam, 1976, p. 163). Although it may not be advisable for researchers to engage in speculation as such, it is healthy to step back every once in a while — if not one hour a day — and consider some of those fundamental issues that rigorous and specialized research all too often forces us to put aside. Accordingly, in this chapter we shall stop and look at the basic question, "Why should children in preschool or in the first years of elementary school be exposed to science?" Based on existing research literature, we shall attempt to formulate a set of explicit justifications for science education in early childhood.

For high school students or young adults, it tends to be easier to find explicit justifications for science education. No doubt, this is because the possibility of a scientific career begins to be imminent for students of this age — and because this is the age when students themselves ask for justifications of all sorts! Gerald Holton, for example, gives these reasons why students nearing or beginning university studies (and not necessarily bound to choose a scientific career) ought to be exposed to science:

> . . . to serve as basic cultural background; to permit career-based opportunities for conceptual or methodological overlap; to make one less gullible and hence able to make more intelligent decisions as a citizen an parent where science is involved; and last but not least, to make one truly sane (for while scientific knowledge is no guarantor of sanity, the absence of knowledge of how the world works and of one's own place in an orderly, noncapricious cosmos is precisely a threat to the sanity of the most sensitive persons). (Holton, 1975, p. 102)

These are perfectly valid reasons, and we agree with them; however, for the most part, they are grown-up reasons. One might argue, of course, that reasons such as Holton's are the true justifications for studying science, and that young children should be exposed to science only to get an early start on the path towards fulfilling those ultimate aims. But this kind of argument only avoids the question. Our task is to find reasons that truly fit young children — not grown-up reasons — reasons which will allow educators to feel that in exposing four, five, six, seven, or eight-year-olds to science they are really doing the right thing. Needless to say, how teachers feel about science is not to be belittled. Several studies in science education refers to elementary school teachers' negative attitudes towards science (Gustafson and Rowell, 1995; McDuffie, 2001; Parker and Spink, 1997; Skamp and Mueller, 2001; Stepans

[1] This chapter appeared as a separate article: Eshach, H. and Fried, M. N. (2005). Should science be taught in early childhood? *Journal of Science Education and Technology*, 14: 315–336.

and McCormack, 1985; Tosun, 2000; Yates and Chandler, 2001); such attitudes can only be reinforced, if not caused, by a sense that science teaching in early childhood may at bottom be a merely nugatory exercise.

In pursuing our goal, we shall proceed in this chapter as follows. First, we consider two basic justifications of science education that science is about the real world and that science develops thinking. Although in the end we do not reject these claims, we do show that, by themselves, they are fraught with difficulty and need to be qualified. With these qualifications in mind as well as research pertaining to children's cognitive abilities, inclinations, conceptions and misconceptions, we present in the second part of this chapter, our own explicit justifications for science educations in early childhood. Finally, we consider some particular learning situations in line with the justifications set out in the second part.

SCIENCE AND TWO BASIC JUSTIFICATIONS FOR SCIENCE EDUCATION

As a term, 'science' is used to describe both a body of knowledge and the activities that give rise to that knowledge (Zimmerman, 2000); whether justified or not, one generally refers to accounts of atoms, forces, and chemical processes as well as one of observing, measuring, calculating as 'scientific'. Science indeed may be thought of as comprising two types of knowledge: *domain-specific knowledge*, and *domain-general-knowledge strategies or domain-general strategies skills* (Zimmerman, 2000). Domain- specific knowledge refers to the knowledge of a variety of concepts in the different domains of science. Domain-general knowledge refers to general skills involved in experimental design and evidence evaluation. Such skills include observing, asking questions, hypothesizing, designing controlled experiments, using appropriate apparatus, measuring, recording data, representing data by means of tables, graphs, diagrams, *etc*., interpreting data, choosing and applying appropriate statistical tools to analyze data, and formulating theories or models (Keys, 1994; Schauble *et al*., 1995; Zimmerman, 2000). The division between domain-specific and domain-general knowledge mirrors other analogous and well-known distinctions, for example, that between conceptual and procedural knowledge, especially in its most general formulation as the division between 'knowing that' and 'knowing how to' (e.g. Ryle, 1949).

This division in the use of the word 'science' and the kinds of knowledge it embraces corresponds to the two main justifications science teachers often use to argue that students as young as preschool should be exposed to science:

1. Science is about the real world.
2. Science develops reasoning skills.

The first statement emphasizes, obviously, domain-specific or conceptual knowledge: by understanding scientific concepts in specific domains children might better interpret and understand the world in which they live. The second statement emphasizes domain-general or procedural knowledge: 'doing science', it claims, contributes to the development of general skills required not only in one specific domain, but also in a wide variety of domains, not necessarily scientific ones.

These two justifications are hardly new; they have accompanied the development of science education tenaciously since the 19th century. Reformers in England, such as Richard Dawes and James Kay-Shuttleworth in the mid-19th century, stressed in their defense of science education the importance of 'useful knowledge' and of 'teaching the science of common things' (see Layton, 1973, esp. chapter 5); students, in other words, should study science because through it they learn about their own world, about the things around them. On the other side of the divide, stood figures such as John Stevens Henslow (better known because of his influence on the young Charles Darwin). Henslow was a botanist and thought of systematic botany as a model subject for science education; he did so, however, not because of its intrinsic interest but because it was, for him, an ideal vehicle for learning observation, exercising memory, strengthening critical thinking, and so on (Layton, 1973, chapter 3). T. H. Huxley, too, belonged to Henslow's camp, and his much-quoted statement that "Science is nothing but trained and organized common sense" (Huxley, 1893, p. 45) summarizes the credo that science should be taught because, in some general way, it helps form powerful ways of thinking.

That science is about the real world and that it develops reasoning both seem even now reasonable enough claims — at least as much so as the division in scientific knowledge from which they are derived. But though teachers continue to use these claims as justifications for teaching science to children, historians and philosophers of science, and scholars in science education as well, have shown them to be problematic and needing qualification. Let us, therefore, take a brief look at the difficulties with these two basic justifications.

Is Science about the Real World?

Driver and Bell (1986) accept that science, in some sense, is about the world. They also argue that "it is about a great deal more than that. It is about the ideas, concepts and theories used to interpret the world." Einstein and Infeld have stated this position famously as follows:

> Science is not just a collection of laws, a catalogue of facts it is the creation of the human mind with its freely invented ideas and concepts. Physical theories try to form a picture of reality and to establish its connections with the wide world of sense impressions. (Einstein and Infeld, 1938)

Thus, one cannot say, simply, that science is 'about the world' for, as the Einstein–Infeld quotation suggests, one must distinguish between a world of 'sense impressions' and a world of 'ideas and concepts' (Driver and Bell, 1986). And, far from what Popper liked to call the 'Baconian myth' (Popper, 1963), abstracting facts into concepts or theories does not follow from simple observation and experiences in the world. On the contrary, according to Schwab and Brandwein (1966), the conceptions and ideas created by the human mind have much to do with how we observe and experience the world: "It tells us what facts to look for in the research. It tells us what meaning to assign these facts" (p. 12).

Consider the following example (the reader may find another example in Driver and Bell (1986)): A child gently kicks a block on the floor so that the block moves forward a little. The sense impression of this 'real world' experience includes the block and its motion, the floor, and the child that we can see. However, the explanation of the case involves the concepts of force, mass, friction, velocity, and acceleration — but none of these is immediately observable; none belongs to the world of our senses or can be abstracted in any direct way from it. Physics concepts like force and mass guide our observations; they tell us what to look for. Thus, only after one comprehends concepts such as velocity, acceleration, and force does one interpret and describe the block's behavior in those terms.

It is not surprising, therefore, that research on science education in the last three decades provides ample evidence that both students *and teachers* hold misconceptions in various domains (Newtonian mechanics: Clement, 1982, 1987; McCloskey, 1983; Electricity: Cohen *et al.*, 1983; Geometrical optics: Galili and Hazan, 2000; Guesne, 1985). For example, in relation to the previously presented example, it is well documented in the literature (Halloun and Hestenses, 1985) that most students believe mistakenly that the 'kicking force' still exists and continues to act on the block even after the boy's foot has left it. As to why the block eventually stops, most students will explain that this is because the force acting on it finally 'runs out'. These ideas, of course, are consistent with the quasi-Aristotelian notion held by many students that where there is motion there is a force producing it (McCloskey, 1983; Viennot, 1979). Accounting for the 'simple' real world occurrence, the kicking of the block, requires the understanding of abstract concepts and principles. Moreover, even those who understand the relevant concepts and principles may find it difficult to apply them in this kind of 'real world' case. Understanding scientific concepts is not an easy task even for many adults. Indeed, Wolpert, in his book on *The Unnatural Nature of Science* (1992), makes the point that, "Scientific ideas are, with rare exceptions, counter-intuitive: they cannot be acquired by simple inspection of phenomena and are often outside everyday experience . . . doing science requires a conscious awareness of the pitfalls of 'natural' thinking" (Wolpert, 1992, p. xi).

To summarize, it is true that science allows one to see the world, but it does so through its own special concepts. Thus, Driver, Guesne, and Tiberghien say that, "In teaching science we are leading pupils to 'see' phenomena and experimental situations in particular ways; to learn to wear scientists' 'conceptual spectacles' " (Driver *et al.*, 1985, p. 193). But if science is more than what we experience directly with our senses, if it is somehow an 'unnatural' activity, as Wolpert says, and if understanding scientific concepts and applying them in specific 'real world' situations is difficult even for adults, we need to ask even more urgently, "Should young children indeed be exposed to scientific concepts?" Perhaps, we should wait until they are more mature intellectually and more able to handle scientific ideas. Moreover, researchers have shown that ideas which take shape in early childhood do not readily disappear with age, but prove to be disconcertingly robust (Black and Harlen, 1993; Gardner, 1999). Should we worry then, that by exposing children to science before they possess the cognitive ability to cope with science, we might, unwittingly, cause misconceptions to take root, which will be hard to undo later on in school, rather than preventing them?

We shall return later to the problem of children's conceptions and misconceptions and then to the questions above. But for now let us just keep them in mind and consider the second basic justification for science education, namely, that science education might contribute to the development of scientific reasoning.

Does Science Develop Reasoning Skills?

At the heart of scientific reasoning both within and outside of professional science is the coordination of theory and evidence (Kuhn and Pearsall, 2000). Taken by themselves, knowledge of theory and knowledge of evidence, naturally, are instances of domain-specific knowledge. From the last section, however, it is clear that science is not science where there is no pairing between theory and evidence. But the coordination of theory and evidence involves inquiry skills or domain-general knowledge, and for this reason, inquiry is considered inherent to science. Science education is thought to contribute to the development of scientific reasoning, accordingly, by engaging students in inquiry situations. This is the view expressed by Chan, Burtis, and Bereiter when they say that in formulating questions, accessing and interpreting evidence, and coordinating it with theories, students are believed to develop the intellectual skills that will enable them to construct new knowledge (Chan *et al.*, 1997).

This same view, which has firm historical roots, is also well documented in educational reports as playing a part in setting modern policy for science teaching. Moreover, such reports have emphasized the importance of developing scientific reasoning *in all age groups*. Here are two examples:

1. According to the report of the Superior Committee on Science, Mathematics and Technology Education in Israel ('Tomorrow 98'), it is extremely important to establish "patterns of investigative thinking as early as pre-school" (1992, p. 26).
2. The Science as Inquiry Standards of the National Science Education Standards (NSES) also advocates that "students at all grade levels and in every domain of science, should have the opportunity to use scientific inquiry and develop the ability to think and act in ways associated with inquiry, including asking questions, planning and conducting investigations, using appropriate tools and techniques to gather data, thinking critically and logically about relationships between evidence and explanations, constructing and analyzing alternative explanations, and communicating scientific arguments" (NSES, 1996).

Literature on scientific reasoning, however, suggests that there are significant strategic weaknesses which have implications for inquiry activity (Klahr, 2000; Klahr *et al.*, 1993; Kuhn *et al.*, 1988, 1992, 1995; Schauble, 1990, 1996). According to Kuhn *et al.* (2000),

> . . . the skills required to engage effectively in typical forms of inquiry learning cannot be assumed to be in place by early adolescence. If students are to investigate, analyze, and accurately represent a multivariable system, they must be able to conceptualize multiple variables additively coacting on an outcome. Our results indicate that many young adolescents find a model of multivariable causality challenging. Correspondingly, the strategies they exhibit for accessing, examining, and interpreting evidence pertinent to such a model are far from optimal. (p. 515)

It seems that there is a gap between the belief that science education based on inquiry will promote scientific reasoning and the reality that students may not have the cognitive skills necessary to engage in inquiry. If even young adolescents, not to mention adults, lack these cognitive skills, surely we cannot expect them in kindergarten and first year elementary school students. But if this is the case, can we expect that young children will benefit from science education based on inquiry? And can we expect young children, then, to develop the kind scientific reasoning that is supposed to arise from inquiry?

Considering the tremendous amount of money, manpower and time required to develop science curricula and prepare teachers to teach them, questions such as these (which taken together constitute the proposal counter to ours, namely, that science should *not* be taught to young children) cannot be taken lightly. This chapter does not presume to give conclusive answers to the difficulties raised in the last two sections. Even so, we do believe it is vitally important to keep such difficulties in mind so that justifications for science education — including those which we shall presently describe — be adopted soberly and with a degree of caution. That said, we think justifications *can* be given for exposing young children to science that at least make taking up the enterprise more reasonable than rejecting it. To this, then, we now turn.

SIX REASONS FOR EXPOSING YOUNG CHILDREN TO SCIENCE

In this section, we consider six reasons as to why even small children should be exposed to science. These reasons are:

1. Children naturally enjoy observing and thinking about nature.
2. Exposing students to science develops positive attitudes towards science.
3. Early exposure to scientific phenomena leads to better understanding of the scientific concepts studied later in a formal way.
4. The use of scientifically informed language at an early age influences the eventual development of scientific concepts.
5. Children can understand scientific concepts and reason scientifically.
6. Science is an efficient means for developing scientific thinking.

Before we describe each reason in detail, two remarks must be made. First, these six reasons are not completely independent of one another. For example, the third, fourth, fifth, and sixth reasons are clearly interrelated. Second, as we stated in the introduction, we are not opposed to the two basic justifications for science education discussed in the last section even though we recognize the difficulties related to them. Thus, our fifth and sixth reasons are completely in line with the general claim "Science develops reasoning skills," and our third and fourth reasons with the claim, "Science is about the real world." However, the way our justifications are formulated avoids, to a great degree, the problems in the traditional justifications, as we shall see, and certainly gives the teacher reasons for science education relevant specifically to young children.

Children Naturally Enjoy Observing and Thinking About Nature

Aristotle began his work the *Metaphysics* by saying, "All men by nature desire to know. An indication of this is the delight we take in our senses . . ." (*Metaph.* 980a, trans. R. D. Ross) (Aristotle, 1941). Aristotle does not use the words 'by nature' (*kata physin*) lightly; for him, the desire to know, even when misguided, is very much at the heart of what it means to be a human being. And he knows that the expression of this natural desire is found not just in the learned discussions of university researchers, but also, as he says, in the mere "delight we take in our senses." This desire to know is not limited to adults. "From birth onward, humans, in their healthiest states, are active, inquisitive, curious, and playful creatures, displaying a ubiquitous readiness to learn and explore, and they do not require extraneous incentives to do so" (Ryan and Deci, 2000, p. 56). In other words, from childhood onwards, humans have intrinsic motivation to know. By *intrinsic motivation* we mean, *doing* an activity for its inherent satisfactions rather than for some separable consequence. Indeed, research on children's motivation to learn and their under-achievement reveals that young children are full of curiosity and a passion for learning (Raffini, 1993). When we recognize this we recognize that children's enjoyment of nature — their running after butterflies, pressing flowers, collecting shells at the beach, picking up pretty stones — is also an expression of their basic desire and intrinsic motivation to know. Conversely, we see that children's knowing and learning about nature, indeed our own knowing and learning too, is a kind of openness to an engagement with nature.

Is the children's involvement with nature, however, in any way intellectual, that is, can it be related to science? Are not children just playing? Yes, they are, but as Vygotsky, among others, has made clear to us, playing is in fact very serious business; play is, for Vygotsky, a central locus for the development of relationships between objects, meanings, and imagination (e.g. Vygotsky, 1933/1978). The pleasure children take in nature, in playing, in collecting, and in observing, make them, in this way, temperamentally ready not only for the things of science but also for their first steps toward the ideas of science.

But what makes young children particularly ready for science is their sense of wonder and intrinsic motivation, and for the educator, this is one of the most important arguments for including science. Educators must work thoughtfully to preserve that sense of wonder, which is so much directed towards the natural world and natural phenomena. In a beautiful essay entitled *The Sense of Wonder* — which, though non-academic, really should be required reading for all future science educators! — Rachel Carson makes the case as follows:

> A child's world is fresh and new and beautiful, full of wonder and excitement. It is our misfortune that for most of us that clear-eyed vision, that true instinct for what is beautiful and awe-inspiring, is dimmed and even lost before we reach adulthood. If I had influence with the good fairy who is supposed to preside over the christening of all children I should ask that her gift to each child in the world be a sense of wonder so indestructible that it would last throughout life, as an unfailing antidote against the boredom and disenchantments of later years, the sterile preoccupation with things that are artificial, the alienation from the sources of our strength.

If a child is to keep alive his inborn sense of wonder without any such gift from the fairies, he needs the companionship of at least one adult who can share it, rediscovering with him the joy, excitement and mystery of the world we live in. (Carson, 1984, pp. 42–45)

So, the first reason why young children should be exposed to science is that, on the one hand, they are already looking at the things with which science is concerned and already in the way the *best* scientists do, i.e. with a sense of wonder; but on the other hand, children are in danger of losing their interest and their sense of wonder if we fail to tend to them and nourish them in this regard.

We said that children are already predisposed to learning about the things of science. It is worthwhile to look at another direction, i.e. that the world offers them sufficient material to feed their interest. Not only the natural world but also the world constructed by human beings with the help of science, which imposes itself upon children. Most parents know, sometimes to their chagrin, that, say, a toy telephone will not hold a child's attention the way a real telephone will. Children are easily absorbed by turning a switch and watching a light go on and off. Bicycle wheels, radios, power tools, lenses and prisms, are all fascinating objects which apply and reflect scientific understanding.

As we discussed earlier, however, the way science ultimately allows us to see the world is by providing us with concepts with which we can frame its phenomena — and it was because these concepts are not always simple or obvious that we questioned wisdom of teaching science to young children. When we consider the remaining justifications we shall reexamine the ability of science education to introduce scientific concepts to young children. But before that, it is important to say that even before concepts come fully into play there is room for *mere looking*, for mere paying attention to phenomena in the world. Such mere looking too is essential to science; indeed, Cesere Cremonini and Giulio Libri's *refusal* to look through Galileo's telescope in 1611 (Drake, 1978, pp. 162–165) still epitomizes an *anti*-scientific spirit.

The world possesses many fascinations, and children, as we said, are taken with them when they see them; often though they need to be led in the right direction. This is where science education is important in children's early years. By pointing and asking questions, with no further explanation, teachers can help children find an abundance of objects and phenomena that will later give content to important scientific concepts (a process about which we shall have more to say below). A teacher often does greater service by simply pointing at the heart-shaped curve of light reflected in a cup of milk than by speaking about the concept of a caustic, or by showing how a comb will deflect a stream a water after the comb has been run through one's hair than by speaking about static electricity, or by asking a child why the merry-go-round keeps turning after it has been pushed than by trying to explain the concept of inertia.

Of course, mere looking requires what one might call 'disciplined openness' — the ability to resist premature explanations. So while the richness of interesting phenomena in the everyday world is a reason to expose young children to science, it remains a challenge for teachers (and for science education to help them) to separate the exposure to phenomena from the interpretation of it. The failure to make that separation in teachers' own minds, moreover, is one reason they might hesitate to expose very young children to science, fearing ineluctable misconceptions. But although the danger of misconceptions is real, as we have said and will emphasize again, well designed science education can help students look while maintaining the openness needed to crystallize the scientific concepts which will ultimately allow them a different, more refined, way of looking at the world.

Exposing Students to Science Develops Positive Attitudes towards Science

Although children have a predisposition to explore the world around them, exposing them to science activities might enhance their motivation and further their natural interest. In addition, we claim that exposing children to science might also inculcate positive attitudes towards science. The term *attitudes* has a variety of meanings. However, according to Miller *et al.* (1961), there are several points of consensus: (1) that attitudes are feelings, either for something or against it; that they involve a continuum of acceptance (accept–reject, favorable–unfavorable, positive–negative); (2) that they are held by individuals; (3) that they may be held in common by different individuals; (4) that they are held in varying degrees (there is neither black nor white, only shades of grey between extremes); and (5) that they influence action. For the educator, what is most important is that attitudes influence motivation and interest (Miller, 1961). Bruce *et al.* (1997), summarizing the literature, argue, moreover, that positive attitudes toward any school subject are related to achievement, may enhance cognitive development directly, and will encourage lifelong learning of the subject in question, both formally and informally. Attitudes towards science classes also have been found to be the best predictors of students' later intentions to enroll in science classes (Crawley and Black, 1992).

It is clear that development of attitudes toward science begins early (Bruce *et al.*, 1997). Lin (1994) found that as early as kindergarten, children's attitudes toward science and their participation in it, were strongly defined. If attitudes are already formed at early stages of life, and if they indeed have significant influence on the child's future development, educators should build environments in which students will enjoy science and have positive experiences connected with it.

Early Exposure to Scientific Phenomena Leads to Better Understanding of the Scientific Concepts Studied Later in a Formal Way

Through experience in everyday life, even when very young, we acquire knowledge about things. We do not only acquire experience and store it but rather organize it. We identify categories of things, dogs for example, in part to avoid having to remember every single dog we have ever seen. Thus, our knowledge is organized to help us decrease the amount of information we must learn, perceive, remember, and recognize. For this reason, Collins and Quillian (1969) aptly called organizational principle, 'cognitive economy'. This economy facilitates reusing previous knowledge structures when possible. This means that general concepts, for example, the concept 'cat', in this view, are treated as efficiently organized information. According to

Heit (1997), perhaps the most dramatic example of concept learning is the performance of young children, who can learn up to 15,000 new words for things by the age of six (Carey, 1978). Admittedly, knowing the word 'cat', say, and knowing the concept *cat* are two different achievements; they are, nevertheless, closely related (Clark, 1983). Concepts consist of verbal as well as non-verbal knowledge representations, including information in the various sensory modalities (Paivio, 1986; Kosslyn, 1994). The concept 'cat', then, not only consists of verbal information such as 'a cat is an animal with four legs, fur, *etc.*', but also, *visual* information — an image of the cat; *haptic* information — we may remember the feeling of a touch of a cat; *aural* information — every one can repeat the *miao* sound of the cat; *olfactory* information — we might even bring in the smell of a cat (especially those who have cats).

Learning a new category is greatly influenced by and dependent on one's previous knowledge and what one knows about other related categories (Heit, 1997). Thus Ausubel could write:

> If I had to reduce all of the educational psychology to just one principle, I would say this: The most important single factor influencing learning is what the learner already knows. Ascertain this and teach him accordingly. (Ausubel, 1968, Epigraph)

More specifically, Heit (1994) points out that the learning of new categories involves the integration of prior knowledge with new observations. According to him, the initial representation of a new category is based on prior knowledge and is updated gradually as new observations are made. This is consistent with *constructivist* perspectives, where one of the main tenets is that learning, construction of novel understandings, and making sense of new experiences are built on prior existing ideas that learners may hold (Driver and Bell, 1986).

Thus, it stands to reason that early exposure to science-related activities with rich verbal and non-verbal information will lead to the formation of deep reservoirs of material which, little by little, may become organized into rich concepts. Negative and sad evidence for this, of course, is the poverty of scientific concepts among students whose childhood was spent in poor socio-economic environments. Indeed, according to Lee (1999) cultural funds of knowledge, brought from students' home lives, provide a basis for making sense of what happens at school and constitute the building blocks on which new knowledge can grow. Students from upper-middle and upper-class families possess a cultural advantage for achieving school-related success that lower-class students do not (Bourdie, 1992; Sahlins, 1976; Wills, 1977).

But since the child's world is full of things related to science anyway, as we said above, it would seem that no special effort has to be made to ensure that children encounter scientific phenomena and that early exposure to scientific phenomena, therefore, need not be an issue for science education. We would argue, however, that *how* children are brought to such phenomena must be pursued with care; we must make sure that while the exposure to scientific phenomena be rich, it should not be capricious. This is because children will begin the process of organizing their experiences into concepts whether we like it or not, and everything they are exposed to will come into play, one way or another. It is not surprising, then, that research has found that novices' concepts are often different from the accepted scientific concepts. Furthermore, these preconceived notions may be inadequate for explaining observable scientific phenomena (Bodner, 1986; Cho *et al.*, 1985; Sanger and Greenbowe, 1997) and may produce systematic patterns of errors (Smith *et al.*, 1993). Such conceptions of students have been labeled by a wide variety of terms in the literature, including misconceptions, preconceptions (Clement, 1982), alternative conceptions (Hewson and Hewson, 1984), and naïve beliefs (McCloskey *et al.*, 1980). According to Smith *et al.* (1993), these terms all indicate fundamental differences between novices and experts. But such terms also indicate the fact we have been emphasizing here, namely, the simple fact that whether they are misconceiving or preconceiving, children are ever engaged in forming ideas about the world.

This last fact, which is the foundation of the constructionist vision of learning, suggests that processes of learning, construction of novel understandings, and sense of new experiences are all ongoing and all influenced by and built on learners' prior existing ideas. Whatever misconceptions children have acquired, then, will also guide their subsequent reasoning. It has been found, moreover, that those misconceptions may be deep-seated and resistant to change (McCloskey, 1983). Designing learning environments in which young children are exposed in a paced and controlled way to scientific phenomena, may help children organize their experiences to be better prepared to understand the scientific concepts that they will learn more formally in the future.

The Use of Scientifically Informed Language at an Early Age Influences the Eventual Development of Scientific Concepts

In previous sections we stressed paced and thoughtful exposure to scientific phenomena as a way to guide the eventual formation of scientific concepts; in other words, the reasons we gave for exposing young children to science always placed scientific concepts in the future. But if there is any truth in what we said at the beginning of this chapter, exposing children to science cannot be so easily divorced from exposing children directly to scientific concepts. What this means is that while 'mere looking', as we stressed above, is essential to science, exposing young children to science requires also justifying "talking" science, that is, using scientific concepts. The question here is, in a way, the opposite of that in the last section: here we need to ask not how experience will help develop scientific concepts but how introducing scientific concepts may influence how children see the world. However, one should also be aware that language and prior knowledge are strongly related to one another. Language, as we shall show, contributes to the formation of the prior knowledge. In this sense, this section is a continuation of the previous one.

The question of how introducing scientific concepts may influence how children see the world, in more general terms, is the question of how language and intellectual development interact. There have been, as Boyle (1971) points out, three traditional schools of thought: the Russian school, dominated by Vygotsky, saw language as the principal mediator of all higher mental functions (see Vygotsky, 1934/1986) and, therefore, as virtually a *sine qua non* of mental growth; the Genevan school under Piaget saw intellectual development as a more or less biological process which is neither initiated nor sustained by language but which is certainly reflected in the child's use of language; the Harvard school, taking more of a middle of the road approach, regarded language as a valuable tool employed by individuals in shaping their experience.

Our general theoretical outlook leans towards the moderate position of the Harvard school. To begin with, it is clear that experience with science, not necessarily verbal, can be extended and can enrich other experiences, helping children to look at phenomena which they might otherwise have ignored. It is also clear that language facilitates this process. Consider a child who played with pulleys in her kindergarten. Now imagine that the child went with her parents on a skiing trip and rode on a ski lift there. Being exposed in the kindergarten to pulleys increases the chances that the child will notice that there are pulleys in the lift system. She might now talk to her parents about the pulleys and might even tell the kindergarten teacher that she saw pulleys in a ski lift. Being exposed to pulleys in the kindergarten prepared the child to notice the pulleys which she probably would have ignored otherwise about them allowed her kindergarten experience to enter into her after-school experience and then her after-school experience to go back again to that in her kindergarten.

The way experience and our understanding of experience can influence language has been observed by Galili and Hazan (2000) in connection to optical phenomena. They argue that language, historically, was developed under the influence of visual perception well before our present understanding of vision was reached. As a result, many linguistic constructions do not conform to present-day scientific knowledge and may lead to student misconceptions. Phrases in our daily language such as "throw a glance" or "give a look," in the authors' view, are probably related to the ancient, and incorrect, Empedoclean idea that vision involves the emission rather than reception of light by the eyes. In a similar manner, Eshach (2003) has shown that the way we talk about shadows in our daily lives may also reveal a strong association between language and ideas regarding shadows. We talk about shadow as an existing entity, e.g., "look at my frightening shadow," "my shadow follows me," and so on. Such phrases may lead students, and adults as well, to attribute the properties of material substances to shadows, rather than understanding them merely as the absence of light. The influence of language might also explain why many students think that "when two shadows overlap, one may diffuse into the other"; similarly, the use of the word 'ray' rather than, say, 'flux,' may be related to students' misconception that there is nothing between the light rays, so that as the distance increases, the area of "nothing" increases and, as a result, a bigger diffused shadow will be created (Eshach, 2003). Just as a particular understanding of optical phenomena may influence language, language can also shape the way one thinks about optical phenomena.

A further example of how language can affect experience comes from investigations concerning students' understanding of sound (Eshach and Schwartz, 2004). All the students in the authors' research used the phrase 'sound waves' when explaining sound. The authors argued that it is apparent that most students' mental image of sound is similar to that of water waves. They believe that sound is a type of matter that travels through water in a sine-wave-like pattern moving up and down. Thus, during the interviews most of the students used up, down, and forward hand movements to describe how sound travels. In day-to-day language, the term 'wave' is commonly used in reference to sound, i.e., 'sound waves.' When describing voices as 'waves,' physicists actually mean that the change in the medium pressure (solid, liquid, or gas) may be expressed as a wave function. The term wave has nothing to do with the shape of the 'voice trajectory path'. The apparently correct expression 'sound waves' used in day-to-day language is interpreted literally, rather than conceptually. As a result, people mistakenly associate sound waves with water waves.

How language influences science-related thinking is strikingly apparent in multi-cultural study such as that carried out by Hatano *et al.* (1993) concerning children's ideas of the concept *living*. In English, the one term *living* is sufficient to distinguish living and non-living things. In Hebrew, however, there are three basic terms relating to living and non-living things — plants, dead objects, and animals. Comparing American, Japaneese, and Israeli students, Hatano *et al.* (1993) provided kindergarten, grade 2, and grade 4 students with lists of items including humans, animals, plants, and various other inanimate objects. The students were asked to categorize the items in the list as *living* or *non-living*. They were also asked questions related to these categories e.g., Can this thing die? or Can this thing grow? The authors found that, for example, only 60% of the Israeli students categorized plants as 'living things' whereas almost 100% of the American and the Japanese students did so. The authors argued that these differences stem from the differences between the Hebrew and English languages, noting that in Hebrew there is a strong association between the term 'animal' and 'living' which does not exist between 'plant' and 'living' (in Hebrew, animals and only animals are called, literally, 'life-owners'). Moreover, while in English one verb, 'to grow', suffices for both plants and animals (including human beings), in Hebrew, there is one verb for animals and a separate verb for plants. Similarly, while in English one says, equally, that a plant, an animal, or a human being 'dies', in Hebrew, there are distinct terms for plants and animals.

These examples not only make clear the power of language to shape experience but also how conflicts can occur between everyday language and scientific language. It is part of a scientist's education to get over these conflicts; but should it be a part of a child's education as well? Should we perhaps avoid scientific language with children, and encourage only everyday language? Would this not still leave room for language's facilitating role in extending and enriching children's experience with scientific phenomena, as in the example of the pulleys and ski-lift? Would it not be better to keep scientific concepts for the future? Our view is that to avoid the tension between everyday language and scientific language and, thereby to avoid possible misunderstandings and misconceptions is to misunderstand how that tension is essential in the learning of scientific concepts. Here we agree with Vygotsky when he writes that "to introduce a new concept means just to start the process of its appropriation. Deliberate introduction of new concepts does not preclude spontaneous development, but rather charts the new paths for it" (Vygotsky, 1934/1986, p. 152). For Vygotsky, the introduction of scientific concepts sets off a process in which the scientific concept reaches downward to the child's everyday or spontaneous concepts while the child's everyday understanding reaches upward to the scientific concept (Vygotsky, 1934/1986, pp. 194–195; it is in this context, incidentally, that Vygotsky introduces his famous 'zone of proximal development'); the tension created is only a sign that this process is underway.

Another advantage of using scientific language as early as childhood lies in the idea that conversations might also influence how one thinks. According to Sfard (2000) "what happens in a conversation along the interpersonal channel is indicative of what might be taking place in the 'individual heads' as well." In other words, the mechanism of thinking, according to the author, is "somehow subordinate to that of communication." Thus Sfard can say, "Both thinking and conversation processes are *dialogical* in character: Thinking, like conversation between two people, involves turn-taking, asking questions and giving answers, and building each new utterance — whether audible or silent, whether in words or in other symbols — on previous ones in such a manner that all are interconnected in an essential way." This at least suggests that if we expose children to 'science talk' it will help them to establish a pattern of 'scientific conversations' which might assist in developing patterns of what we call 'scientific thinking'. As Brown and Campione (1994) put it:

> It is essential that a community of discourse be established early on in which constructive discussion, questioning and criticism are the mode rather than the expectation. Speech activities involving increasingly scientific methods of thinking, such as conjecture, speculation, evidence and proof become part of the common voice of the community. (Brown and Campione, 1994, p. 229)

To create such a community of discourse in the classroom, teachers may first simply be aware of the influence of language on the reception, internalization, and comprehension of scientific concepts and prepare themselves accordingly. Subsequently, they may actively include phrases in their discussions with the students that encourage discourse — simple phrases such as, "How do we know?" "Let's hypothesize," "What do you think may happen if . . . ?" "How did we get to that conclusion?" "Let's check," "How can we check?" (More specific and fuller examples of how appropriate language may be used to promote scientific understanding are presented in the section, "Some learning situations — language and prior knowledge").

Children Can Understand Scientific Concepts and Reason Scientifically

Earlier in this chapter, we discussed how concepts or theories, which are not the result of mere direct experience of the world with our senses, are often hard to understand, even by adults. Does this still stand as an objection to what we have just been arguing? Is there any evidence that children are indeed able to deal with scientific concepts, that is, that they are sufficiently mature intellectually to comprehend scientific concepts? This question is still crucial. We agree that: (1) children naturally enjoy observing and thinking about nature; (2) exposing children to science develops positive attitudes toward science; (3) early exposure to scientific phenomena leads to better understanding of the scientific concepts studied later in a formal way; and (4) the use of scientifically informed language at an early age influences the eventual development of scientific concepts. But if children are not mature enough to think scientifically, if they are not mature enough to understand scientific concepts (which are often subtle and sometimes complicated) can we truly gain much from exposing them to science?

True, scientific concepts may be hard to grasp even by adults; however, this does not mean that children *cannot* think abstractly about scientific concepts. On the contrary, literature shows that children are able to think about even complex concepts.

Metz (1995), for instance, critiques the assumption that children at the concrete operational level are 'concrete thinkers,' whose logical thought is linked to manipulation of concrete objects. This assumption is supposedly derived from Piaget's work, but Metz argues that a close look at Piaget's writings themselves give little evidence that this is what Piaget truly thought. She claims that Piaget did indeed believe that school children's thinking is directed towards some concrete referent, but not that the product of their thinking is concrete. According to Metz, Piaget's writings reveal numerous examples of abstract constructs which were formulated, at least on an intuitive level, by elementary school children; these include speed (Piaget, 1946), time (Piaget, 1927/1969), necessity (Piaget, 1983/1987), number (Piaget *et al.*, 1941/1952), and chance (Piaget and Inhelder, 1951/1975). One specific example provided by Metz (1995) is the case of cardinal numbers. Piaget *et al.* (1941/1952), Metz (ibid) believed that children develop an understanding of cardinal number, an idea that clearly transcends the concrete, around 7 or 8 years of age. Even earlier, between 6 and 8 years of age, Piaget claimed that children come to construct the idea of chance, in the sense of the "nondeductible character of isolated and fortuitous transformations" (Piaget and Inhelder, 1941/1975, p. 214).

Another objection to what we have been arguing in the previous sections may arise from our earlier discussion of science education based on inquiry, namely, that the gap between the belief that science education, based on inquiry, will promote scientific reasoning, and the reality according to which even young adolescents may not possess the cognitive skills necessary to engage in inquiry (Kuhn's *et al.*, 2000). Kuhn's *et al.* (2000) conclusion, in this regard, concurs with early cognitive development research (Dunbar and Klahr, 1989; Inhelder and Piaget, 1958; Kuhn *et al.*, 1988; Schauble, 1990). These researchers suggested that before the age of about 11 to 12 years children have very little insight into how hypotheses are supported, or contradicted by evidence, and that even at this age, and into adulthood, understanding is quite shaky (Ruffman *et al.*, 1993).

Other research, however, shows that even younger children show the ability to think scientifically. For instance, Gelman and Markman (1986) showed that 4-year-old subjects could appropriately select surface information or deeper natural-kind membership information to form inductions, depending on the question asked. Ann Brown's (1990) study of 1-to-3-year-olds exploring simple mechanisms of physical causality documented that toddlers reasoned from deep structural principles, as opposed to surface features, when they had access to deeper information. Ruffman *et al.* (1993) showed that already by 5 years of age children may distinguish between a conclusive and an inconclusive test of a hypothesis.

There are several explanations for the differences of opinion in the research community as to whether or not small children can think scientifically. For instance, Sodian *et al.* (1991), criticizing Kuhn *et al.* (1988) pointed out that: (1) The tasks discussed included contexts in which children had strongly-held beliefs of their own – it is very plausible that revising such beliefs is more difficult than forming theories when no prior beliefs exist or when beliefs are not held with any degree of conviction – and (2) The tasks were too complex. Consequently, according to Sodian *et al.* (1991), Kuhn *et al's* research tended to underestimate children's understanding of hypothesis-evidence distinction.

We wish to present another problematic issue concerning these kinds of research. Although cognitive development studies refer to "scientific thinking," "scientific reasoning," or "scientific discovery," as the processes by which children explore, propose hypotheses via experimentation, and acquire new knowledge in the form of revised hypotheses, these studies are sometimes carried out in *non-scientific contexts*. Such studies use what Zimmerman (2000) calls *simulated discovery tasks* method. Three examples demonstrate this point:

Example 1: In a study by Kuhn *et al.* (1988) described in their book *The Development of Scientific Thinking Skills*, children were told that the type of cake eaten — either chocolate or carrot — affected whether or not persons caught colds. Children were then given access to evidence — i.e., they were shown who ate which cake and who went on to catch a cold. They were then asked to explain how the evidence showed the relevance of particular variables, to say which variables were casual, and to conclude which hypothesis was correct. The authors found that when asked to assess the evidence children either ignored the evidence and insisted that it was consistent with their prior theories, or they used the evidence to construct a new theory but failed to grasp that this new theory contradicted their previously-held theory.

Example 2: In the study, "Reflecting on Scientific Thinking: Children's Understanding of the Hypothesis-Evidence Relation" (Ruffman *et al.*, 1993, experiment 1), four-year-old children were introduced to an imaginary character named Sally. Sally was then said to have gone off to a playground where she could no longer see or hear anything happening near the children. The children were then shown drawings of five boys eating either green (or red) food and had several teeth missing, and another group of drawings of five boys eating red (or green) food who possessed a complete set of strong and healthy teeth. For half the children green food was associated with tooth loss and for the other half, red food was associated with tooth loss. All children associated the correct food with teeth loss, showing that they had no difficulty in interpreting the covariance evidence. The experimenter then 'faked' the evidence by rearranging the 10 pieces of food so that it now appeared that opposite food was the source of tooth loss. With this, Sally 'returned' and observed the evidence; the children were asked to say what kind of food she would say cause kids' teeth to fall out. The children were required, thus, not only to form the correct hypotheses themselves, but also to understand how the evidence might lead Sally to form a different hypothesis. The authors found that five-year-old children and even some four-year-old children understood the hypothesis–evidence relation.

Example 3: Sodian *et al.* (1991) told children a story about a big mouse or a small mouse living in a house. They were then shown two boxes, each with a piece of cheese inside, and were told that the mouse would eat the cheese if it could. One box had a large opening wide enough for either mouse; the other box had a small opening wide enough only for the small mouse. The children were asked which box they should use to determine whether there is small or big mouse in the house. Children recognized that to determine the size of the mouse it was better to set out the box with the small opening.

In all three examples, children's ability to coordinate evidence with hypotheses was investigated in non-scientific contexts; no scientific concepts were required for the tasks given to the children. While such research contributes tremendously to our understanding of how children connect hypotheses to evidence, it must also be admitted that considering scientific reasoning, without engaging in science, might provide only an incomplete and inadequate picture of scientific reasoning processes. The tendency to separate scientific reasoning from science may, in fact, be related to the lack of communication between cognitive developmentalists and science educators (Strauss, 1998). Strauss (1998), with whose view we concur, writes, "Developmentalists often avoid studying the growth of children's understanding of science concepts that are taught in school" (p. 358).

To summarize, assuming children are able to understand complex concepts and are able, to some extent, to connect theory and evidence, educators should, in our view, expose children to situations in which those abilities may find fertile ground to grow. In the next section, we shall consider such situations more closely and adduce positive arguments for learning scientific reasoning skills in specifically scientific contexts.

Science is an Efficient Means for Developing Scientific Thinking

At first glance, this statement seems blatantly tautological and, therefore, useless as a reason to justify teaching science. Yet, the issue is more subtle than it appears. What goes by the name 'scientific reasoning' or 'scientific thinking' covers more ground than what goes by the name 'science' alone. At the same time, the kind of thinking that real scientists engage in is not necessarily what one likes to call 'scientific'. Let us say a little more about these two points.

First, as we described at the beginning of the chapter, science comprises both domain-specific knowledge and domain-general knowledge. In view of this, scientific reasoning, scientific thinking, or scientific discovery include both conceptual and procedural aspects. The conceptual aspects of scientific thinking are inseparable from scientific content domains; however, the procedural aspects can easily break away from content. It is these procedural aspects that we tend to have in mind when we speak about scientific thinking as analytical or critical thinking or, especially, thinking which connects evidence and theory. In this sense, it can be said that we employ scientific reasoning in our daily lives even when the subject is not science! This is probably the justification for the research, described in the previous section, that investigated so called scientific reasoning in non-scientific contexts.

Having said that, one must be careful about going too far: calling every instance of reasoning, every instance of connecting evidence and theory, as scientific. Consider the following two examples.

1. Before going to school, John left a new toy, which he had just received for his birthday, on the desk in his room. When he returned from home, the first thing he wanted to do was to play with the toy. But when he went to get it, he discovered it was not where he left it. His parents, as far as he knew, were still at work so, there was no one to ask: he had to solve the mystery himself. How might he proceed? First, he makes some hypotheses: (1) there was a thief in the house who stole the toy; (2) one of his parents got back early from work and moved it; (3) his sister, who usually comes home from school before John, took the toy to a friend of hers. Having set out these hypotheses, he can now examine them one by one. Regarding the first, he can check whether any of the windows are open or broken, whether the back door is open or whether there is anything else missing from the house. To test the second hypothesis, he can check whether one of his parents' bags is in the house or some other personal belongings indicating that one of them had arrived before John came home from school. As for the last hypothesis, he can look for signs showing that his sister was already home. For instance, he can check whether or not her room is tidy and arranged as it was in the morning.

2. A different kind of example in which it might be said that evidence and theory are brought together is this. Based on evidence from their intelligence services, several world governments, the American and British governments chief among them, constructed a theory that Iraq under Sadam Hussein's regime had illicit weapons of mass destruction threatening America, Britain, and other parts of the world. They decided, therefore, to launch a war on Iraq and replace Sadam's regime. The public too is involved in deliberations concerning the war and, to the extent that this is an issue in the presidential election, will have to make a judgment in the end. Based on reports in the media, citizens gather data and form and test different hypotheses. They might weigh new evidence showing the extent to Sadam's cruelty, discoveries of mass graves, evidence of horrific torture, and so on, and join this evidence with a theory justifying the removal of nasty leaders by anyone who has the power to do so.

Both examples show how the idea of scientific thinking can be pushed too far. Nevertheless, they do bear some marks of genuine scientific reasoning: in the first case, for example, there is the discovery of an anomaly (John's toy not being where he expected it to be), and, in both, hypotheses are formulated and subsequently tested by looking for evidence, evidence is coordinated with the hypotheses, and perhaps, new hypotheses are formed. The second example diverges from scientific thinking most clearly in that both the governments involved and the voting public are weighing evidence not against a theory of how things are, but against what is perceived to be a *desirable* course of action, that is, their reasoning occurs within a value system, not a conceptual system. The fallacy of assuming that this is a *scientific* process was pointed out long ago by G. E. Moore (1903), and it is still a fallacy committed by many engaged in social or political issues. The ways in which the first example diverges from scientific thinking are less obvious. The main problem, though, is that while there are hypotheses there is no theory, that is, no overarching view of how things are. There is no attempt to " 'recognize where on the map' a particular object of study belongs" (Toulmin, 1960, p. 105); hypotheses alone do not make a theory, even a simple-minded one. It is important to realize how such cases diverge from scientific thinking because, otherwise, it becomes all too easy to conclude that science is unnecessary for developing scientific reasoning.

Such examples could conceivably be used to develop those elements of scientific reasoning which they do indeed contain: one can learn through them to formulate hypotheses in a sensible way, and one can learn to be critical. But then one would have to be careful to bring out the divergences which we just described. Learning to recognize such divergences would, of course, not be a bad thing, but it could not be done without some other model examples of scientific thinking. Pursuing scientific thinking in this way, then, would prove to be a cumbersome and unduly complicated affair. Our view, thus, is that while it is not impossible to use non-science examples to develop scientific thinking, it is more *efficient* to use one from science.

Take for instance, an investigation of the influence of light on plants; it is rich in domain-general knowledge. First one must identify the relevant variables: the light, the soil type, the amount of water, the temperature, the humidity, and plant species. Then to examine the influence of light, children can design a set of experiments in which all the variables are kept constant except for the light. They can check for changes in the degree or rate of growth, color alterations, light-induced movements (*phototropisms*), and so on. Seeing sets of experiments where only one change is allowed to occur focuses children's attention on the meaning of variables and control variable; they can reflect on the problems which can arise by altering more than one variable; they form hypotheses and suggest ways of testing them; they see how one hypothesis may lead to another. Moreover, they can repeat the experiment to examine the influence of other variables.

Thinking in this context exposes children to 'clean' situations where they can (sometimes even immediately) see the influence of an isolated variable, as opposed to complex situations where there are many variables and no easy way to control them. Having this kind of experience, then, children are likely to be better prepared to see that even in a 'simple' situation such as that of John's toy, one can not control or isolate the variables. For instance, the open window doesn't necessarily mean that there was a burglar — it might be that the sister and not the burglar opened the window. This is true *a fortiori* with regard to the Iraq example where even the task of identifying the variables is formidable! Thus, by beginning with scientific thinking in scientific contexts — and one ought not forget that the *model* for scientific thinking in any context still comes from science! — children not only learn to be critical and analytical but also learn to see more easily and clearly where other kinds of thinking fails to be 'scientific'.

What it means to be or to fail to be ‘scientific’ is a question teachers must ask themselves continuously; students even the very young ones we are speaking about, ought to begin to ask as well. Popper’s ideas, although in other respects outmoded (and we shall have more to say about this in a moment), are still a good starting point for asking what it means to be scientific. Using scientific contexts to develop scientific thinking is also the ideal way to introduce the Popperian view of science. According to Popper (1959) a theory is scientific only if it is falsifiable, that is, if it is such that one indubitable counter-instance refutes the whole theory. Furthermore, while a genuine scientific theory, in Popper’s view, can be tested and falsified, it can never be incontrovertibly verified. Neither the most rigorous tests nor the test of time shows a theory to be true; a theory can only receive a high measure of corroboration and may be provisionally retained as the best available theory, until it is finally falsified (if indeed it is ever falsified) or is superseded by a better theory.

An example such as the following does well to illustrate these ideas. Consider the following situation: two objects, one heavier than the other, are released from the same height. According to the Aristotelian theory, the objects will reach the ground in an amount of time inversely proportional to their masses. So, for instance, if the mass of one object is twice that of another then it will fall to the ground from the same height in half the time. Now, let’s think of the following two experiments:

Experiment 1:

Release a feather and a stone from the same height (Fig. 1). It will be observed that the stone will reach the ground faster. Thus, the experiment apparently proves Aristotle’s theory that heavier objects, if released from the same height, will reach the ground faster than lighter objects.

Experiment 2:

Repeat experiment 1, but this time use a sheet of paper instead of a feather (Fig. 2). Again, the Aristotelian theory holds true. “Is there any need to go on?” the teacher might ask. Let us perform a third experiment:

Experiment 3:

Release two stones, one heavier than the other, from the same height (Fig. 3). Let the stones fall onto a hard surface so that one can hear when they hit the surface. It will

Figure 1. A feather and a stone released from the same height.

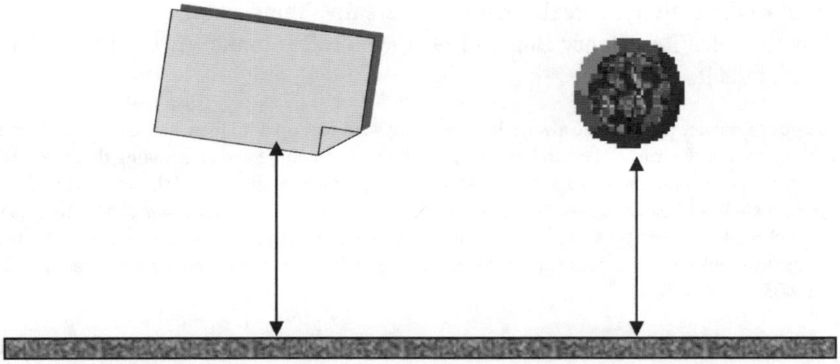

Figure 2. A sheet of paper and a stone released from the same height.

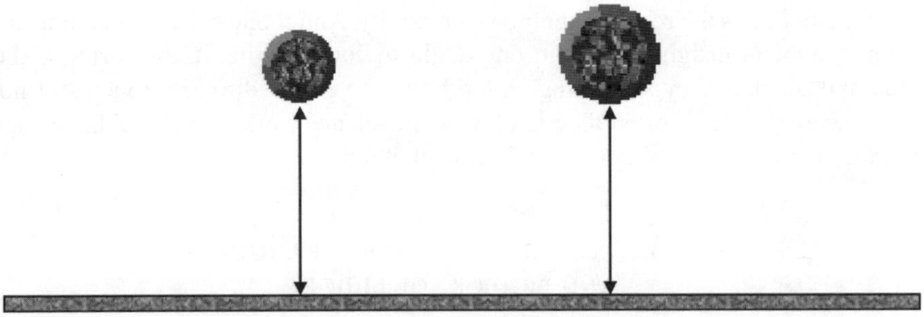

Figure 3. Two stones released from the same height.

be observed that the stones reach the ground at the same moment (and the sound of the two stones hitting the surface will, consequently, be heard simultaneously).

Experiment 3 falsifies Aristotle's theory, even though that theory was considered true for over a thousand years, and even though other experiments *were* consistent with what Aristotle thought. Through this example, then, one easily sees how positive experiments are always at best tentative, and therefore, the scientific theories they are meant to demonstrate must be viewed as tentative as well. This is much more difficult to show in non-scientific contexts. In the example John's toy, for instance, there are too many hypotheses which can all be easily contradicted; the idea of 'falsification' in that kind of non-scientific context becomes highly problematic.

Moving away from this basically Popperian view of science, investigation such as that concerning the influence of light on plants or the falling objects also brings out the second point we made at the start of this section, namely, that the kind of thinking real scientists engage in is not always what one likes to call 'scientific'. For quite some time already, the preoccupation of historians and philosophers of science (Kuhn, Polanyi, Feyerabend, etc.) has shifted from a fixed notion of a 'scientific method' to the activity of real scientists as creative thinkers who do not necessarily 'follow the rules'. As Henry Bauer (1994), who refers to the 'myth of the scientific method', puts it:

> The corpus of science at any stage always includes only what has, up until then, stood the test of time. We see nothing in it of the trial and error, backing and filling, dismantling and rearranging that actually took place in the past, be that centuries ago or just a few years ago. Only when we read the actual accounts written by early studies of nature do we begin to realize how many errors and false starts there were that left no traces in modern scientific texts. Once can give excellent, objective, rational grounds now for the science in the textbooks, but that does not mean that it was actually assembled in an impartial, rational, steady manner. (Bauer, 1994, p. 36)

It is only by being involved actively in thinking about something so 'objective' as the influence of light on a plant that one can gain this insight into how science really works. Children will begin to have a hint that, for example, asking whether a plant will be induced to move by light is not a question dictated by any perfectly determined method; it is the result of their own creativity. And if one believes that this kind of 'philosophical insight' can wait, one ought to consider that in the cartoons they watch and pictures they see young children will be exposed to *other* views of how science works — more often than not a view of science working in a cold, mechanical, inhuman way, according to an inflexible method.

SOME LEARNING SITUATIONS — LANGUAGE AND PRIOR KNOWLEDGE

Here we provide a selection of learning situations connected with specific scientific concepts, to provide concrete illustrations of some of the ideas we have been discussing, particularly, how language and prior knowledge may influence the development of scientific concepts.

Heat and Temperature

Many children conceive 'cold' as the equal counterpart to 'hot', instead of understanding 'cold' and 'hot' in terms of the absence or presence of heat. This misconception is well demonstrated in children's answers to the following question:

"Given two cups, one metal and the other foam, which cup will keep a cold drink cold for longer time? Which cup will keep a hot drink hot for longer time?"

Many students mistakenly believe that a metal cup will keep the drink cold for longer time and the foam cup will keep the hot drink longer. One reason many students give for their answer is that cold drinks (like coke) are usually kept in metal cans while coffee is usually served in foam cups to keep it warm. These answers indicate that students separate 'coldness' from 'hotness' as independent qualities, and, it may be surmised, students do so because of their prior everyday experience with hot and cold things.

Simple experiments with young children may be conducted to show that a foam cup or a thermos keeps *both* hot and cold drinks longer. We believe that such experiments may lead children to understand that the same isolated container can keep hot drinks hot and cold ones cold, though we do not think they will necessarily grasp immediately the precise scientific ideas involved. On the other hand, as we have been claiming throughout, these experiences are likely to make children better prepared to grasp the scientific ideas later.

Optics

Many students believe that shadows are material entities. Feher and Rice (1988) found that nearly 50% of their research participants believed that shadows exist in the darkness, so that a dog, for example, would still have a shadow when it walked into the full shadow of a house. Some participants thought that light was necessary only to illuminate the shadow (as if it were just another object), whereas others believed that light actually caused the shadow's visibility (*e.g.* by heating it up). Galili and Hazan (2000) found that 9th-grade students (pre-instruction students), 10th-grade students (post-instruction students), and college students (teachers college) regarded shadows as things which can be manipulated as independent objects and can be added or subtracted. They also understood shadows to be things which remain randomly oriented in space, regardless of any light source, that the shadow of the object represents its shape much as its mirror image does, and that light merely "makes [a shadow] visible." In fact, shadows are reified (as in Feher and Rice, (1988)) like images in mirrors and lenses. Langley *et al.* (1997) found that most 10th-grade students, before formal instruction, drew light rays that rarely extended as far as the shadow. The authors argued that this indicated that students failed to understand the relationship between light propagation and shadow formation.

It is likely that children will more easily come to understand that a shadow is not an entity itself, if teachers, *already in preschool*, associate shadows with the absence of light rather than the presence of some definite thing. It might help to provide explanations such as this: "You see all around the area of shadow there is light. In the shadow area there is no light (or less light in the case of several light sources)". But since, as we mentioned above, these ideas about shadows may derive from the language used to describe them (Eshach, 2003), teachers can take advantage of language in playful ways to challenge children's ideas: besides phrases such as "a shadow follows me" they can say, for example, "a spot of 'no-light' follows me."

Archimedes' Law of Buoyancy

The usual answer as to why certain objects float is that they are lighter than the water. Most of students do not grasp that it is the relationship between the relative densities of the object and the water that determines whether or not the object will float, and not their relative weights.

Density is considered a difficult concept for children. Yet, teachers can demonstrate the idea of density for kindergarten children, in such a way as: the teacher fills a container with water and asks what happens if one drops a small stone in the water. Children will generally say that the stone will sink because it is heavier than water.

The stone does sink, but is it really heavier than the water? To check, the teacher places the stone on one side of the balance scale and the water, removed from the container and transferred to a plastic bag, on the other. Seeing that the water is heavier than the stone, the students must face the fact that the stone sinks even though it is lighter than the water. From here, the teacher places the stone inside a balloon without inflating it, ties it so that no water can get inside, and asks what will happen to the stone with the balloon if we put them inside the water. The balloon with the stone will sink. However, if we inflate the balloon while the stone is inside, the stone-balloon combination will float. The experiment is effective because the weight variable is kept, more or less, constant (in fact, of course, the weight *increases* slightly!) while the volume changes dramatically. Exposing children to the possibility that not only the weight of an object, but also its volume, may determine whether or not an object sinks or floats, paves the way, we believe, to the concept of density and will make it easier to grasp when introduced formally in student's later studies.

Newton's Third Law

Consider the following question: Two children, Sharon and Ruth, sit in identical wheeled office chairs facing each other. Sharon places her bare feet on student Ruth's knees. Sharon then suddenly pushes outward with her feet. The following three situations should be presented (possibly by using different pairs of children) each at a time: (1) Sharon is bigger than Ruth; (2) Sharon is smaller than Ruth; and (3) Sharon and Ruth are of the same size. Who moves when Sharon pushes outwards with her feet, Sharon or Ruth? Explain the answer. Obviously, according to Newton's third law, both will move (though with accelerations depending inversely on their masses) since the force Sharon's feet exert on Ruth equals the force Ruth's knees exert on Sharon. Yet, many young students believe that whoever is bigger, or is the one actively pushing, must exert a greater force, that is, the bigger or active person is somehow the 'more forceful' person. According to Hestenes *et al.* (1992), this belief stems from the way people interpret the idea of 'interaction'. They often use the 'conflict metaphor' according to which the 'victory belongs to the stronger'. Thus the more active, heavier, or bigger 'wins' in the 'struggle'; they 'overcome' their 'opponent' with a greater force.

Sharon and Ruth, by being the active agents, as it were, in the experiment described above, have a good chance of realizing that in an interaction between objects not only the stronger exerts a force but that there is a force acting on both objects. It is not our intention, of course, to teach Newton's Third Law to kindergarten children. However, with the right teacher's help, we believe that such experiments where children actually feel the forces at work can help to make the Third Law, which is notoriously difficult to grasp, seem natural and intuitive when it is studied later on.

SUMMARY AND CONCLUDING DISCUSSION

In this chapter, we stepped back and considered the question, "Why should children in preschool or in the first years of elementary school be exposed to science?" Let us review the main points of the chapter.

We began by looking at the two basic justifications most often used by educators for why preschool students should be exposed to science, namely, that science is about the real world, and that science develops reasoning skills. Though we did not reject these justifications, we tried to bring out the problematic aspects of them that make it difficult to accept them *tout court*.

Regarding the claim that *science is about the real world*, we showed that science is not about the world in a *direct* way; it, in a sense, is about a great deal more than the world. For one, by abstracting facts into concepts or theories, scientific insights do not follow from simple observation and experiences in the world. Nor are scientific concepts always evident in the way ordinary appearances are — a fact reflected in the difficulty even adults have in grasping scientific concepts. On the other hand, we see the world with the help of conceptions and ideas created by the human mind; they are like glasses that help us be aware of things to which we might otherwise be blind. But this also means that there is a danger of putting on inappropriate glasses that distort our vision. With such glasses, then, children might develop misconceptions that may be difficult to undo later.

As for the claim that *science develops reasoning skills*, we showed that it is not clear that the preconditions for this are always fulfilled. In this connection, we cited literature showing that even young adolescents, not to mention young children, lack the skills required to engage effectively in many of the forms of inquiry necessary for the first steps in scientific reasoning. Engaging children in tasks requiring investigation might bring them only frustration.

In both cases, one is left with the serious question, "Should young children who may not yet be mature to intellectually handle scientific concepts and scientific inquiry indeed be exposed to science?" Should we take the risk of introducing science to young children, when, as a result, they might develop misconceptions, which may be hard to change later?

With those concerns on the table, we tried to reformulate the arguments for exposing young children to science, so that, in the balance, educators might feel that there are better reasons for teaching science to young children than withholding it from them. The arguments and some of their normative implications, in brief, were as follows:

1. Children naturally enjoy observing and thinking about nature. Whether we introduce children to science or whether we do not, children are doing science. We are born with an intrinsic motivation to explore the world. This means that children will be taking their first step towards science with or without our help. To prevent missteps, it is wise to intervene and provide learning environments that will be conducive to children's developing, in a fruitful way, a scientific outlook and assimilating material for learning scientific concepts later.

2. Exposing students to science develops positive attitudes towards science. Attitudes are formed early in childhood and can have crucial impact on children's choices and successes in learning science. If we wish for our children to develop positive attitudes towards science we must introduce science in a way that will pique their curiosity and spur their enthusiasm.

3. Early exposure to scientific phenomena leads to better understanding of the scientific concepts studied later in a formal way. Prior experience has significant influence on the development of new knowledge. This is a reason for scientific education, with the aid of a sensitive teacher, because how children are brought to such scientific phenomena must be pursued with care; we must assure that while the exposure to scientific phenomena be rich, it should not be capricious.

4. The use of scientifically informed language at an early age influences the eventual development of scientific concepts. Language has a significant influence on concept construction. Sometimes, however, conflicts can arise between everyday language and scientific language. But, following Vygotsky, we argued that these kinds of conflicts and tensions, if accompanied by thoughtful science-educational practice, can be the source of genuine concept development. Approaching the question of language from a different direction, we also argued that the connection between mechanism of thinking and that of communication suggests that exposing children to 'science talk' will help them to establish patterns of 'scientific conversations' which, in turn, might assist in developing patterns of 'scientific thinking'.

5. Children can reason scientifically. Although some research has shown that children lack the requisite skills to conduct investigations fruitfully, other research has shown that children as young as 4 years old, can, nevertheless, distinguish between a conclusive and inconclusive test for a hypothesis. If children have, thus, the seeds of skills that allow them to connect theory and evidence, it is reasonable that exposing them to situations where they can exercise these skills will further develop them. These situations must be planned in advance so that they fit the children's abilities, and in this science education plays its crucial role.

6. Science is an efficient means for developing scientific thinking. By pursuing scientific thinking in scientific contexts children are more easily exposed to 'clean', 'objective' situations where they can see the influence of an isolated variable; children, in this way, not only learn to be critical and analytical, but also learn to see more readily and plainly where other kinds of thinking fails to be 'scientific'.

Ideally, a kindergarten science program would give expression to all six of these themes. But the spirit, at least, of these themes can be found in the preschools of Reggio Emilia, Italy. Referring to a Newsweek article which declared these preschools to be the best in the world, Howard Gardner wrote, "in general I place little stock in such rating, but here I concur" (Gardner, 1999, p. 87). According to Gardner, the Reggio Emilia preschool program is such that groups of children spend several months exploring themes which interest them: sunlight, rainbows, raindrops, shadows, ant colonies, lions' dens, poppy fields, an amusing park for birds built by the youngsters, and fax machines. The children approach these things from many angles; they ponder questions and consider phenomena that arise in the course of their explorations; and they end up creating artful objects that picture their interests and their learning: drawings, paintings, cartoons, charts, photographic series, toy models, and replicas. Thus the children of Regio Emilia are allowed to explore the things of nature and science according to their own desire; they are encouraged to ask questions and find ways to synthesize and formulate their thoughts about what they see; they are surrounded by people who believe that these early experiences in science are far from fruitless. The Reggio Emilia approach will be further discussed in the next chapter.

Windows of Opportunities

In the discussion above, we chose not to emphasize findings from brain science, which some might see as an unforgivable lacuna; nevertheless, one must make choices in such matters! Still, we do not by any means want to imply that brain science ought to be neglected; indeed, it is likely to offer important insights for educational questions in the future. For this reason, we want to close with a few points from those studies that touch the question of whether science should be taught to K-2 children.

In his impressive and insightful book, *The Disciplined Mind*, Howard Gardner (1999) relates how he heard the following pronouncement made by a prominent neuroscientist in a conference:

> This is the decade of the brain. We are going to know what every region of the brain does and how the various part of the brain work together. And once we have attained that knowledge, we will know exactly how to educate every person. (Gardner, 1999, p. 60)

Gardner, who claims that he generally avoids unpleasant exchanges in conferences, said that this speaker had managed to raise his hackles. Extreme statements beget extreme responses, so, at the conclusion of the talk, Gardner retorted:

> I disagree totally. We could know what every neuron does and we would not be one step closer to knowing how to educate our children. (Gardner, 1999, p. 60)

With Gardner, we believe that brain studies will never be able to tell us exactly how we should educate our children. That notwithstanding, it is undeniable that learning has to do with the production of neurons and their interconnections, and, it has been shown that this tremendous productive activity slows down to a close at about the age of 10 (Nash, 1997). To ignore these facts (and Gardner certainly does not!) in considering when and how education should begin thus seems to us to be a grave mistake.

Gardner goes on to say:

> Decisions what to teach, how to teach, when to teach, and even how to teach entail value judgments. Such decisions can never be dictated by knowledge of the brain. After all, if children learn patterns well when they are young, that constitutes equal reason for teaching them math, music, chess, biology, morality, civility, and hundred other things. Why should foreign language get priority? [the case of language was mentioned by the conference speaker who said that according to brain studies it is better to teach children foreign languages at first grades] You can never go directly from knowledge about brain function to what to do in first grade on Monday morning. And the decision one makes about teaching languages might well differ, and properly so, depending on whether you live in Switzerland, Singapore, Iceland, or Ireland." (Gardner, 1999, p. 61)

We completely concur with Gardner that brain science will never determine what exactly we should teach and how we should do it. Our view that we should teach math, music, chess, biology, morality, civility, and a hundred other things; and especially that we should teach those subjects that come under the heading of 'science' is not a deduction from brain science. What we *do* learn from brain research is that, once we have decided that science is important, we may not have all the time in the world to pursue it. In the 1990s, much research was being published showing that leaning in specific domains, where 'learning' is understood as a modification of neural structure, occurs most efficiently within certain 'critical periods' or 'windows of opportunity', and that these 'windows of opportunity' begin to close at around the fourth grade (Nash, 1997; Shore, 1997). The classic case is foreign languages, which tend to be harder and harder to learn as one gets older. For essential science skills, such as logic and mathematics, the window seems to close quite early (Begley, 1996). It is not that one cannot learn later in life, but, as Nash (1997) puts it, "while new synapses continue to form throughout life, and even adults continually refurbish their minds through reading and learning, never again will the brain be able to master new skills so readily or rebound from setbacks so easily" (p. 56).

Of course these findings from brain science, strictly speaking, go against Bruner's famous thesis that "any subject can be taught effectively in some intellectually honest form to any child at *any stage* [emphasis added] of development" (Bruner, 1960, p. 33); however, they do support his statement that subjects, and most of all science, could be taught at a *young* age — indeed, these findings show that science *should* be taught at a young age! It is, therefore, incumbent on the science educator to provide children with environments, materials, and activities, to develop their scientific reasoning while these 'windows of opportunity' are still open. Entering those open windows will prepare children to enter the doors of the society as good citizens possessing the ability to question, to critique, and to learn.

CHAPTER 2

HOW SHOULD SCIENCE BE TAUGHT IN EARLY CHILDHOOD?

Equipped with the six reasons to expose small children to science given in the previous chapter, I now shift from philosophy toward a more pragmatic direction: How should science be taught to children?

I start this chapter with an intriguing story taken from Richard Feynman's charming book *What do you Care What Other People Think* (Feynman, 1989). The story describes how Melville Feynman taught physics to his young child, Richard, during weekend walks through the Catskill Mountain woods. Richard Feynman eventually became a famous, renowned Nobel Laureate in Physics. His father, Melville most likely inadvertently, used rather advanced educational approaches to teach his son. These approaches would undoubtedly have been rare in the schools of those times. I will use this story as a framework to discuss and develop several distinct educational approaches which I believe provide insight into science education in early childhood.

HOW RICHARD'S FATHER TAUGHT HIS SON SCIENCE

"On weekends, my father would take me for walks in the woods and he'd tell me about interesting things that were going on in the woods . . ."

"One kid says to me, "See that bird? What kind of bird is that?"

I said, "I haven't the slightest idea what kind of bird it is."

He says, "It's a brown-throated thrush. Your father doesn't teach you anything!

But it was the opposite. He had already taught me: "See that bird?" he says. "It's a Spencer's warbler." (I knew he didn't know the real name.) "Well, in Italian, it's a *Chutto Lapittida*. In Portuguese, it's a *Bom da Peida*. In Chinese, it's a *Chung-long-tah*, and in Japanese, it's *Katano Tekeda*. You can know the name of that bird in all the languages of the world, but when you're finished, you will know absolutely nothing whatever about the bird. You will only know about humans in different places and what they call the bird. So let's look at the bird and see what it's *doing* — that's what counts." (I learned very early the difference between knowing the name of something and knowing something.)

He said, "For example, look: the bird pecks at its feathers all the time. See it walking around, pecking at its feathers?"

"Yeah."

He says, "Why do you think birds peck at their feathers?"

I said, "Well, maybe they mess up their feathers when they fly, so they're packing them in order to straighten them out."

"All right," he says. "If that were the case, then they would peck a lot just after they've been flying. Then, after they've been on the ground a while, they wouldn't peck so much any more — you know what I mean?"

"Yeah."

He says, "Let's look and see if they peck more just after they land." (Richard P. Feyman, 1989)

It wasn't hard to tell: there was not much difference between the birds that had been walking around a bit and those that had just landed. So I said, "I give up. Why does a bird peck at its feathers?"

"Because there are lice bothering it," he says. "The lice eat flakes of protein that come off its feathers."

He continued, "Each louse has some waxy stuff on its legs, and little mites eat that. The mites don't digest it perfectly, so they emit from their rear ends a sugar-like material, in which bacteria grow."

Finally he says, "So you see, everywhere there's a source of food, there's some form of life that finds it."

Now, I knew that it may not have been exactly a louse, that it might not be exactly true that the louse's legs have mites. That story was probably incorrect in *detail*, but what he was telling me was right in *principle*. (Feynman, 1988, pp. 3–4)

The Teaching Avenue of Feynman's Story: A Summary

1. Identifying a problem to be investigated — "Why do birds peck at their feathers?"
2. Making a hypothesis — "Well, maybe they mess up their feathers when they fly, so they're pecking them in order to straighten them out."
3. Making predictions derived from the hypothesis — according to the hypothesis one may expect that birds peck at their feathers more just after landing than after being on the ground for a while.
4. Designing an experiment — identifying a variable that can (1) be measured, and (2) test the prediction. In this case the variable is the "*pecking frequency*."
5. Collecting data — after deciding upon the variable, the measurements are achievable: comparing the differences between the pecking frequencies of birds which had just landed with those which were on the ground for a while.
6. Obtaining results — Richard and his father found that there was no difference in the frequencies.
7. Interpreting the data and arriving at the appropriate conclusions — based on the findings, they reached the conclusion that birds do not peck at their feathers in order to straighten them after flying.
8. Providing the scientific explanation — Richard's father taught him the *principle* that wherever there's a source of food, there's some form of life that finds it.

As an educator, I would say that it might have been better to encourage Richard to provide more hypotheses and to test them as well. However, there is no doubt that while Richard might not have learned the bird's name, he definitely learned something about the nature of science and gained a better sense of what scientific inquiry means. Moreover, he probably understood the *principle* that his father taught him.

The story illustrates quite well the following educational topics: (1) inquiry-based teaching; (2) learning through authentic problems; (3) preference for the *psychological* rather than the *logical* order; (4) scaffolding; and (5) situated learning. After discussing these in detail, I will review further educational topics that should also be kept in mind in teaching K-2 and beyond; These are: (6) learning through projects; and (7) non-verbal knowledge.

From Factual Knowledge to Inquiry Skills

Richard's father taught his son that learning about things should go beyond knowing their names. To make his point clear, Melville, in a sense of good humor, named the bird in different languages. He pointed out that by knowing the name one can only know about humans in different places and what they would call the bird. Richard learned early in life, as he himself writes, "the difference between knowing the name of something and knowing something." I take the term "name" not literally, but rather, as a symbol of factual-knowledge-based teaching. Schwab *et al.* (1966) calls such teaching — *teaching as a dogma*. According to such teaching, the body of a doctrine is conveyed as absolute truths.

According to Perkins (1992) there is ample evidence demonstrating that schools, which predominantly teach by rote, barely succeed in getting students to acquire knowledge, even at the level of mere memorization, let alone achieve a clear and satisfactory understanding of this knowledge. Melville Feynman had good intuition and understood well what has now become clear in many countries: the aim of science teaching should not only be the teaching of accepted content in science (scientific knowledge), but should also provide children with an understanding of the characteristics and procedures of scientific inquiry (Kanari and Millar, 2004). "Inquiry learning is defined as an educational activity in which individually or collectively investigate a set of phenomena — virtual or real — and draw conclusions about it" (Kuhn *et al.*, 2000, pp. 496–497). This importance of inquiry learning is well supported by many educational reports worldwide. For instance, one standard of the National Science Education Standards (NRC, 1996) is the Science as Inquiry Standards, which "highlight the ability to conduct inquiry and develop understanding about scientific inquiry" (p. 105). The need to teach science as inquiry is also important for the reason which Schwab wrote about in 1966,

> the operations required of our elites to meet our present problems are no longer capable of being understood by the public which has had only a dogmatic education. . . . The problems we now face cannot be solved within the bounds of existing doctrines. These problems require new conceptions and fresh doctrines. These fresh doctrines and conceptions can be acquired only by a course of enquiry proceeding by innovation, trial, and failure. . . . Hence the problem we face: to convey to our publics a view of enquiry, especially of scientific enquiry, which is commensurate with its present character. Otherwise, adequate support and assent will not be given to the enquiries our national problems require. (Schwab *et al.*, 1996, pp. 8–9)

In my view, not only can this be done but sowing the seeds of inquiry skills as early as K-2 science education is crucial. In the classic book *The Teaching of Science* (Schwab and Brandwein, 1966), Schwab states that "an enquiring classroom is one in which the questions asked are not designed primarily to discover whether the student knows the answer but to exemplify to the student the sorts of questions he must ask of the materials he studies and how to find the answers" (p. 67). According to Schwab, learning processes that begin with problems may promote children's inquiry skills. Indeed Feynman's story begins with a problem posed to Richard: "Why do you think birds peck at their feathers?" The problem led to learning, but did not, by any means, test Richard's knowledge. The next section elaborates on the learning through problems approach.

LEARNING THROUGH PROBLEMS

"The ability to solve problems is one of the most important manifestations of human thinking" (Holyoak, 1995, p. 267). "A problem is viewed as a gap between where a person is and where he or she wants to be" (Hayes, 1981). In other words, "a problem arises when we have a goal — a state of affairs that we want to achieve — and it is not immediately apparent how the goal can be attained" (Holyoak, 1995, p. 269). In the preface to the National Association for Research in Science Teaching (NARST) monograph, *Towards a Cognitive-Science Perspective for Scientific Problem Solving*, Lavoie (1995), writes that the monograph "was conceived in response to our Nation's need for a population of scientifically literate individuals who can think and solve problems" (p. iv). He also argues that "a focus on problem solving seems to have taken a back seat, not only in our classrooms, but in our respected science education research circles." He advocates that renewing science educators' focus on problem solving is one of the most important subjects of our research and teaching efforts at all levels. In his call one can identify the latent assumption that with appropriate teaching, educators can assist students in developing their problem solving skills. Although there is ample evidence in cognitive psychology literature that problem solving depends heavily on available specific knowledge pursuant to the domain to which the problem belongs, Holyoak (1995), argues that normal people do acquire considerable competence in solving daily problems. He suggests that problem solving depends on general cognitive abilities that can potentially be applied to an extremely wide range of domains. Taking into account the ideas that educators can help students develop problem solving skills and also that these skills can be used to deal with novel situations, I definitely believe that educators should respond to Lavoie's call. I therefore, present not only the view that problem solving skills can and should be developed as early as childhood, but also provide the means with which to apply this view in K-2 science education.

Returning to Feynman's story, Richard was asked to deal with a problem without previously learning about the subject. Although one might agree that developing problem solving skills is important, there remains the issue of what children can gain from dealing with a question when they do not have the necessary background to answer it. Is it dangerous to allow children to become frustrated? To illustrate my concern I will provide an example of an incident that occurred in China described in Howard Gardner's (1999) excellent book, *The Disciplined Mind*:

> My wife and I were visiting Najing with our eighteen-month-old son, whom we had adopted from Taiwan when he was an infant. Each day we allowed Benjamin to insert the key to the key slot at the registration desk of the Jinling Hotel. He had fun trying, whether or not he succeeded. But I begun to notice that older Chinese people who happened to pass by would help my son place the key in the slot and would look at us disapprovingly, as if to chide us: "Don't you uncultivated parents know how to raise your child? Instead of allowing him to flail about and perhaps become frustrated, you should show him the proper way to do things. (p. 94)

The issue concerning a child's benefit from the aforementioned types of questions is particularly valid in light of certain learning theories that had been embraced until about 25 years ago, which perceived learning as a linear and sequential process (Zohar and Nemet, 2002). Learning was described hierarchically — progress from simple, lower-order cognitive tasks to more complex ones. Bloom (1954) and Gange (1974), Zohar and Nemet (2002) argue that complex understanding was thought to occur only through the accumulation of basic, prerequisite learning.

With regard to Feynman's story, applying such an approach would probably have prevented Melville from starting off with a problem Richard knew nothing about. However, Richard was lucky because it does not seem as if he became frustrated. On the contrary, this event impressed him so much, leaving such a positive feeling with him that even many years later he still remembered and cherished it as he expressed in his book:

> That's the way I was educated by my father, with those kinds of examples and discussions: no pressure — just lovely, interesting discussions. It has motivated me for the rest of my life, and makes me interested in *all* the sciences. (Feynman, p. 4)

In addition, "unlike the older theories, more recent learning theories see learning in a very different way. Rather than evolving from the fragmented knowledge resulting from complex ideas being broken down into smaller parts, understanding is seen as evolving while learners are engaged in thinking and inquiry in contexts that make sense to them" (Zohar and Nemet, 2002, p. 36). Posing an authentic problem that is interesting to a child, despite the fact that the child does not apparently have the necessary background knowledge to deal with it, might be a good starting point for learning. A well known such teaching strategy is termed in literature as *problem-based learning* (PBL). Problem Based Learning started at the Johns Hopkins Medical School in the early 1990s. As mentioned earlier, PBL differs from the traditional approach to teaching, where students first learn the subject matter and only then are given problems as exercises. I will provide some theoretical support as to why the PBL method might be appropriate for children, by describing two types of problems that adults as well as children encounter, namely, *well-defined* and *ill-defined*. I will then discuss two types of reasoning which people naturally utilize when solving problems; *rule-based-reasoning* (RBR) and *case-based reasoning* (CBR). Finally, I will argue that PBL encourages both RBR and CBR, as opposed to traditional teaching methods which neglect CBR. I will thus argue that PBL is an efficient learning environment, which better scaffolds inquiry skills.

Two Types of Problems: Well-Defined and Ill-Defined

There are two types of problems which people may encounter: well-defined and ill-defined problems. To explain the differences between these kinds of problems let us consider Newell and Simon's (1972) view of problem solving as a search in a metaphorical space. According to their theory, the representation of a problem consists of four elements: a description of the initial state at which problem solving begins, a depiction of the goal state to be reached, a set of operators or actions that can be taken, which serve to alter the current state of the problem and path constraints that impose additional conditions on a successful course to solution. The problem space consists of the set of all states that can potentially be reached by applying the available operators. A solution is a sequence of operators that can transform the initial state into the goal state in accordance with the path constraints. The search metaphor is most appropriate when the solver can identify a clear goal, is able to understand the initial state and constraints, and knows exactly what operators might be useful in solving the problem. These cases are considered to be *well-defined* problems. A good example of well-defined problems is a puzzle problem. The knowledge required to solve a puzzle problem is present in the statement of the problem. Children at kindergarten and primary school levels are exposed to many such situations, e.g. games such as chutes-and-ladders, dominos, checkers or chess, all of which are kinds of puzzle problems. The knowledge needed to solve the problem — winning the game — is present in the rules of the game.

However, many daily situations are often poorly defined. Not all the information they require to cope with the problem is present. It is difficult to specify the state from which one can start to identify the operators that might be applicable or even to recognize when the goal has been achieved. In other words, many daily problems are ill-defined in that the representation of one or more of the basic components — the goal, the initial state, the operators or the constraints — are severely compromised. First, let us take a problem that both adults and children face on a daily basis — how to be happy. There is no one way to reach this goal. Moreover, happiness depends on many factors. Things that would make us happy today would not necessarily make us happy tomorrow. Thus, in such a problem even the goal is not defined. Another apparently simple problem a child may face might involve playing in his/her backyard with a ball. While playing, the ball gets stuck in a tree and won't come down. Now, the child has a real problem, especially if this is the first time this has happened. Indeed, what may serve as a problem for one person may be seen as a trivial routine exercise for another (Wheatley, 1995). To cope with this problem the child can call his parents who are inside the house to get the ball for him. Upon realizing that they are busy and will only be able to come later, he has to think of an alternate course of action. He might try to shake the tree. If this still doesn't work the child might try to reach the ball by using a long stick or throwing something at the ball, dislodging it and causing it to fall. The child may also think of bringing a ladder to climb up and reach the ball. In such a problem, unlike in puzzle-like problems, the operators are not defined. The child does not know the operators in advanced and has to invent them.

Despite the fact that most problems in daily life are ill-defined, children at school are primarily given well-defined problems. According to Wheatley (1995), science and mathematics problem solving in school is thought of as the solving of highly structured word problems appearing in texts, aiming at providing practice for prescribed computational procedures. A student can usually decide which method to use by identifying the method illustrated in a preceding lesson. Consequently, one may regard most problems with which children are provided in school as well-defined ones. The author argues that such word problems do not develop students' problem solving skills. One may also find similar situations in early childhood education. Most of the time in school is invested in playing puzzle problems, teaching children to count, solving simple arithmetic problems, or asking the children questions on simple factual knowledge to check whether or not they remember what was learned. Although it is crucially important to develop such skills, it is my understanding that educators should also include ill-defined problems in their lessons. According to

Wheatly (1995), in situations where problem solving is the explicit goal, non routine problems are usually selected.

Another interesting finding about problem solving in school is that developing problem-solving competence does not guarantee conceptual understanding: students may perform well with quantitative problems yet show a severe lack of understanding of the concepts that they are dealing with in these problem solving activities (Mazur, 1992). Furthermore, "several less powerful understandings allow the students to arrive at the 'correct' answer to the physics problem — correct in terms of the expected quantitative solution or algebraic expression" (Bowden *et al.*, 1992, p. 263). Mazur raises the question as to the benefit of mainly teaching the mechanical manipulation of equations without gaining understanding.

So, one question that arises from this discussion is how we, the educators, can promote the development of problem solving of both ill and well defined problems. To better understand what educators face in their efforts to promote children's abilities to deal with both types of problems it is worth understanding the two natural reasoning mechanisms which people employ when dealing with problems: Rule-Based Reasoning and Case-Based Reasoning.

RULE-BASED REASONING

Rule-based reasoning (RBR) is the process of drawing conclusions by linking together generalized rules, starting from scratch (Leake, 1996). RBR models are rooted in the philosophical belief that humans are rational beings and that the laws of logic are the laws of thoughts (Eysenck and Keane, 1995). According to Kolodner (1993), although some rules are very specific, the goal is to formulate rules that are generally applicable. An important advantage of rules in general is the economy of storage they allow (Kolodner, 1993). However, there are some disadvantages to RBR:

- The problem of applicability, i.e., bringing some general piece of knowledge to a particular situation (Mostow, 1983). When rules are expressed too abstractly, the terms tend to be unintelligible to the novice and have a variety of specific meanings to the expert.
- Ill-defined domains. In domains that are not completely understood, the rules do not encompass all of the situations that they are asked to cover or are assumed to cover, may admit tacit exceptions, or can be contradicted and annulled by other rules (Rissland and Skalak, 1991).

These characteristics of rules and RBR indicate that people should use more than RBR when solving puzzle problems, and further, facing authentic daily problems. Let's take, for instance, the game of Checkers. As explained previously, a child playing such a game is actually dealing with a well-defined problem. The goal of the game is either to capture all of the opponent's pieces or to blockade them. Different children may understand the term "capture" or "block" differently. An experienced Checkers player will probably have a broader and better understanding of what the abstract ideas "capture" or "block" mean. Moreover, the same rule might have a different meaning in the game situation. The novice also might find himself, indeed employing the rules, however with no success. When dealing with ill-defined problems this might be even worse. Going back to the ball story, it is clear that, from the child's point of view, unlike in the case of Checkers, there are no rules to which a child can refer. In Richard Feynman's story, Richard faced the same kind of situation. There were no rules available for him. This means that the idea of RBR clashes with the idea of PBL.

An underlying assumption in RBR is that abstract information is important in problem solving, while the value of knowledge of a specific event and specific experiences is neglected. This view is challenged by the personal-knowledge point of view, which views the knowledge of specific episodes as a key to successful problem solving (Cohen, 1996; Kolodner, 1993; Leake, 1996).

CASE-BASED REASONING

Personal knowledge, defined as the unique frame of reference and knowledge of self, is central to the individual's sense of self (Higgs and Titchen, 1995), and is a result of an individual's personal experiences (Butt *et al.*, 1982). Much of the knowledge used in problem solving and making judgments is tacit and individual (Carroll, 1988; Polanyi, 1962).

Case-based reasoning (CBR) takes the idea of personal knowledge one step further. In CBR, the primary knowledge source is not generalized rules or general cases, but a memory of stored cases recording specific prior episodes (Leake, 1996). A case which records knowledge at an operational level represents specific knowledge tied to a context (Kolodner, 1993). Cases may cover large or small time frames, associating solutions with problems, outcomes with situations or both (Koldner, 1993). CBR can mean adapting old solutions to meet new demands, using old cases to explain new situations and using old cases to critique new solutions. It can also require the use of reasoning from precedents to interpret a new situation or to create an equivalent solution to a new problem (Kolodner, 1993). Advantages of CBR include the following:

- It allows the reasoner to propose solutions to problems quickly, saving the time that would be necessary to derive these answers from scratch (Kolodner and Leake, 1996).
- It allows the reasoner to propose solutions in domains that are not completely understood (Kolodner and Leake, 1996). In such situations rules are imperfect. Thus, solutions suggested by cases also increase the quality of the solutions.
- It allows for avoiding making mistakes similar to those made earlier (Cohen, 1996; Kolodner, 1993; Leake, 1996).
- Reference to previous similar situations is often necessary to deal with the complexities of novel situations (Kolodner and Leake, 1996).

In our daily lives, we humans find ourselves confronting ill-defined problems or problems that are not completely understood. CBR assists us in overcoming the complexity of real-life situations. Cases, as opposed to rules, provide a large chunk of knowledge tied to a context. Cases may also contain a wide spectrum of knowledge, including sensory factors that may be ignored by rules. It can be argued that cases, as opposed to rules, provide rich index items and thus may lead to efficient retrieval of relevant knowledge from memory, particularly in ill-defined situations. CBR is especially important in childhood. First, if adults have difficulties in applying some general pieces of knowledge (a rule) to a particular situation, it would most likely be even harder for children to do so. Second, children have not yet been exposed to much in the way of formal learning and thus have not yet acquired rules that might help them in dealing with the rich situations they face in real-life. Children, therefore, are more likely to depend on specific previous cases they have dealt with in their past in order to deal with a new situation. Children who play a new game for the first time will have probably learned from their mistakes in previous specific games that they played before. In the case of the child dealing with the ball in the tree, he might remember another case of a toy stuck on a high shelf or seeing a basketball game where a ball was stuck between the ring and the board. By remembering such specific cases, the child may adapt the solution from the prior case and alter it so that its solution fits the new case.

According to Eshach and Bitterman (2003), the argument that CBR, in many situations, is more efficient than RBR leads to the idea that the recollection of cases, in some situations, is more efficient than the recollection of rules. The authors argue that there are situations where indexing a large chunk of a more specific knowledge (e.g., cases) might result in more efficient retrieval of that knowledge from memory, rather than the retrieval of small pieces of abstract knowledge (e.g., rules). One reason for this, they claim, is that cases record knowledge at an operational level and thus are more meaningful to the reasoner than the abstract knowledge of a rule. In addition, a case, as opposed to rules, provides rich index items and thus may lead to the efficient retrieval of relevant knowledge from memory especially in ill-defined situations.

PBL, as opposed to lecture-based instruction, encourages and promotes CBR. The cases provided by the PBL approach are indexed in memory by rich index items. For example, Richard Feynman may remember the case of walking in Catskill Mountain woods and seeing the birds pecking at their feathers, in a completely different situation. For instance, while learning at school about the connection between food sources and the ability of some life form to find it, or even if he himself was to take his child on walks in the woods. Many routes may lead Richard to the retrieval of this specific story. This, in turn, might assist him in confronting other situations.

To summarize, in a traditional learning environment a child begins to learn a subject by accumulating basic prerequisite rules belonging to that subject. Usually, in this stage only lower-order thinking such as mere memorization is required. Only after acquiring these prerequisites can the learning process progress and allow us to demand a higher order of thinking such as problem solving. More advanced learning environments possess an opposite approach to learning. Within such learning environments, *learning for understanding* occurs when children engage in inquiry that requires higher-order thinking, in contexts that make sense to them. I have also provided an explanation, based on cognitive psychological theories, as to why approaches such as PBL, are efficient. Specifically, I have demonstrated that PBL promotes the use of CBR; a natural reasoning technique which we humans employ throughout daily life. Factual knowledge based teaching, on the other hand, emphasizes mainly RBR which might not be sufficient in dealing with the full complexity of real life situations. Additional support for the use of problems as a starting point for teaching stems from Dewey's distinction between *logical* and *psychological* methods, discussed in the next section.

LOGICAL VS. PSYCHOLOGICAL METHODS

According to Dewey (1916), science is the outcome of observation, reflection, and testing which are deliberately adopted to secure a settled and assured subject matter. He claims that science signifies the realization of the logical implications of any knowledge and that perfecting of knowing, is its final stage. He argues that,

> . . . there is a strong temptation to assume that presenting subject matter in its perfected form provides a royal road to learning. What is more natural than to suppose that the immature can be saved time and energy, and be protected from needless error by commencing where competent inquiries have left off? The outcome is written large in the history of education. Pupils begin their study of science with texts in which the subject is organized into topics according to the order of specialist. (p. 220)

To the non-expert, however, according to Dewey, this perfected form is a stumbling block. Specifically because the material is stated with reference to furthering of knowledge as an end in itself, its connections with the material of everyday life are hidden. Moreover, acquiring the factual rules of a subject does not guarantee the ability to use them precisely when needed. This is due to the characteristics of rules as well as RBR. Dewey, with whom I agree, further argues that "from the standpoint of the learner scientific form is an ideal to be achieved, not a starting point from which to set out" (p. 220). Dewey suggests that the proper way to teach science is to begin with the experience of the learner, with what is familiar to the child and an ordinary acquaintance by him or her. This is a method which he termed the "*psychological method*," or the "*chronological method*."

> Dewy warns us that "Educationally, it has to be noted that logical characteristics of method, since they belong to subject matter which has reached a high degree of intellectual elaboration, are different from the method of the learner — the chronological order of passing from a cruder to a more refined intellectual quality of experience. When this fact is ignored, science is treated as so much bare information, which is less interesting and more remote than ordinary information, being stated in an unusual and technical vocabulary" (p. 230).

Referring to Feynman's story again, one can see that Melville employed the psychological method by beginning the teaching process with an authentic problem that he thought might be of interest to his son. He could "save" time by beginning with the factual rules, and then explaining how the pecking phenomenon can be understood by these rules. However, considering Richard's needs, he understood the necessity of challenging him in order to develop in him an intrinsic motivation and desire to learn.

SCAFFOLDING

Another educational issue that is well demonstrated in Feynman's story is *scaffolding*. The scaffolding metaphor first appeared in Wood *et al*'s (1976) paper *The Role of Tutoring in Problem Solving*. According to this metaphor adults are said to provide a *scaffold*, much like that used by constructors in erecting a building, when their assistance "enables a child or novice to solve a problem, carry out a task or achieve a goal which would be beyond his unassisted efforts" (Wood *et al*., 1976, p. 90). In this paper no explicit reference to Vygotsky's developmental theory (1978) has been made. However, subsequent work, beginning with Cazden (1979), increasingly linked scaffolding with Vygotsky's notion of the zone of proximal development (ZPD). The ZPD is defined as the distance between what individuals can accomplish alone and what they are able to accomplish when assisted by a more capable peer. The increasing use of the scaffolding idea reflects a growing disenchantment with what might be called the individual-child-learner model of development, made popular by followers of Piaget (Stone, 1998). As opposed to Piaget's developmental theory where there is no emphasis on the role of social relationships on the child's accommodation processes, Vygotsky's theory implies that social relationships underlie all higher mental functions. Vygotskian's theory maintains that activities and experiences become internalized only after a series of transformations which initially take place between people (interpsychological) and are then directed inward (intrapsychological), meaning that dialogue with others becomes internalized and part of an individual's inner thoughts (Jones *et al*., 1998).

Wood *et al*. identified six types of assistance which the adult tutor can provide: recruitment of the child's interest, reduction in degrees of freedom, maintaining goal orientation, highlighting critical task features, controlling frustration, and demonstrating an idealized solution path. Stone (1998) suggests that this list includes perceptual components (e.g. highlighting task features); cognitive components (e.g. reducing degrees of freedom); and affective components (e.g. controlling frustration). Other complementary types of assistance that are worthy of mention are those of Carter and Jones (1994): prompting, modeling, explaining, asking leading questions, discussing ideas, providing encouragement, and keeping the attention centered on the learning context. Returning to Feynman's story, one can identify some of the above types of assistance: *recruitment of the child's interest* — Richard's father began by referring to the birds by different names. I believe that this was done with a sense of humor that, in itself, probably invited Richard into the adventure his father was conspiring. In addition, his father declared that knowing the birds' name provides no information whatsoever about the birds. This, I assume, may have increased his son's motivation to begin to wonder about the "real thing," which extends beyond the name. After this motivating introduction there is a direct invitation — So let's look at the bird and see what it's doing — that's what counts. *Reduction in degrees of freedom* — after the invitation to discover what the birds are doing comes a reduction in the degrees of freedom — not to look at all of the bird's behavior, but rather focus only on how it pecks at its feathers. Asking leading questions — after Richard hypothesizes that birds peck at their feathers to straighten them out after flying, Richard's father leads his son to examine this hypothesis by attempting to answer whether or not birds peck more just after they land. Through this teaching process, it is apparent that the father has a clear goal toward which he *maintained orientation* to reach his son the principle that "wherever there's a source of food, there's some form of life that finds it." For this goal, he gave his son with scientific *explanations*.

Use of a good metaphor may help people gain insights about a phenomenon. Indeed, the scaffolding metaphor explicitly reveals the role the teacher takes in the teaching-learning process, which aims at facilitating optimum learning (Fleer, 1992). Teachers should "have an image of scaffolding as a complex social process of communicational exchange and conceptual reorganization through which knowledgeable others foster understandings and capabilities" (Stone, 1998).

SITUATED LEARNING

Situated learning is another educational topic related to Feynman's story. The main tenet of situated learning, which focuses on the relationship between learning and the social situations in which it occurs, is that learning is a process that takes place in a participation framework, as opposed to in an individual mind. As William F. Hanks puts it in his introduction to the Lave and Wenger book *Situated Learning: Legitimate Peripheral Participation*, the individual learner does not gain a discrete body of abstract knowledge which (s)he will then transport and reapply in later contexts. Instead, (s)he acquires the skill to perform by actually engaging in the process, under the attenuated conditions of *legitimate peripheral participation* (LPP). According to LPP, "learners inevitably participate in communities of practitioners and the mastery of knowledge and skill requires newcomers to move toward full participation in the socio-culture practices of a community" (Lave and Wagner, 1991, p. 29). Learning is not the acquisition of knowledge by individuals as much as a process of *social* participation. This contrasts with most traditional classroom teaching that usually involves out of context abstract knowledge. Learning through LPP occurs no matter which educational form provides the context for learning, or regardless of whether there is any intentional education at all. This view point provides a fundamental distinction between learning and intentional instruction. Such decoupling does not deny that learning can take place where there is teaching, but does not take intentional instruction to be by itself the lone source or cause of learning.

Feynman's story is connected to the idea of situated learning because, first, the learning took place in a participation framework — father and son. Second, the learning process was in context, during the actual watching of the birds. Finally, it is reasonable to assume that the learning was unintentional. After all, Richard and his father did not set out on their walk with the purpose of learning about birds, but rather to engage in some quality time together.

So far I have used Feynman's story to describe educational approaches that may fit early childhood science education. Next I will describe some additional educational strategies for those who want to teach science to young children.

LEARNING THROUGH PROJECTS

There are many advantages to problem-based learning (PBL). There are different methods by which one can apply the PBL approach. For instance, Feynman's story is a single short episode activity. Other PBL activities may last longer. The renowned Reggio Emilia preschools in Italy, for example, apply a project-based learning system. In this environment, groups of children spend several *months* exploring a theme of interest such as sunlight, rainbows, raindrops, shadows, the city, ant farms, poppy fields, an amusement park for birds built by the youngsters or the operation of a fax machine. The children approach these objects, themes and environments from many angles. They ponder questions that arise in the course of their explorations and they end up creating artful objects that picture their interests and their learning such as drawings, paintings, cartoons, charts, photographic series, toy models, or replicas. When the exploration of the theme comes to a close the objects that have been created are placed on display so that parents, other children and members of the community can observe them and learn from them. The following example is taken from Gardner (1999):

The Rainbow

Suppose that in the middle of a school day a rainbow appears. Either a child or a teacher notices the rainbow and brings it to the attention of the others. The youngsters begin to talk about the rainbow and, perhaps at the suggestion of a teacher or on their own initiative, a few children begin to sketch it. After the rainbow disappears, the children would probably want to know what happened to it; where it came from and where it went after disappearing. This could well be the first stage in which the children identify both an interesting theme to be explored and are able to derive related and relevant problems to inquire about. In the next stage, the children start collecting data to answer their problems. One of the children might pick up a prism that happens to be nearby and look at the light streaming through it. She might then call over her classmates and they would begin to experiment with other translucent vessels. The next day it rains again, but afterwards the sky is cloudy and no rainbow is visible. Henceforth the children set up observational posts after a storm to guarantee that they will be able to spot the rainbow when it next appears and capture it through various media. If no rainbow appears, or if they fail to capture its appearance, students will confer as to the reasons why and consider how to prepare for better rainbow sighting. This would all mark the beginning of a project on the rainbow. In the following weeks, children gain a common interest in researching rainbows and read and write stories about them, explore raindrops, consider rainbow-like phenomena that accompany lawn hoses and mist, and play with flashlights and candles, noting what happens to the light as it passes through various liquids and vessels. The project does not start off with a specific goal and no one knows where it will eventually land. Also, while previous projects may influence the guidance given by the teacher, this open-ended quality is crucial to the educational milieu that has been established over the decades at Reggio.

Criticism of the Reggio Approach

In the early 1990s, Newsweek declared that the preschools of Reggio were the best in the world. Referring to this declaration, Howard Gardner writes "in general I place little stock in such rating, but here I concur" (p. 87). So the reader may justifiably ask how the author of this book dares to criticize this wonderful approach. I definitely agree that the Reggio approach is unique. However, approaching the topic from the teachers' perspective, I would argue that the projects which might seem sound at first, might not be congruent with the teachers' qualifications and needs. To deal with rainbows, teachers must have decent knowledge about them. They should know about the connection between a rainbow and light streaming through a prism — a very difficult concept for kindergarten teachers to understand, being as they do not usually have sufficient scientific background. I have no doubt that in Reggio this approach works wonderfully. However, if our aim is to expand this approach we might find it difficult to implement. It is my view that teachers who have insufficient background might teach the scientific ideas so terribly wrong that it would be extremely hard to alter at a later stage. In many cases, unlike the superior conditions at Reggio Emilia, a kindergarten teacher usually works alone, without many opportunities to learn from colleagues and experts.

I thus argue that educators should seek after such activities that not only fit the children's needs but also the teachers' abilities, motivations, and needs. To summarize this point, the Reggio Emilia's project-based approach sufficiently considers the child's needs. Moreover, it may even contribute tremendously to the children's cognitive development. But to succeed in using such an approach, a kindergarten teacher must receive sufficient scientific support. Without such support, one may not only miss the approach's goal, but may also unintentionally lead students to misconceptions. Thus, I argue that K-2 science education should be teacher-centered as well as student-centered, as opposed to the traditional student-centered approach. This subject will be further discussed in depth in Chapter 4. Nonetheless, projects chosen with care may fit the kindergarten environment.

In all the above teaching approaches, both verbal and non-verbal representations are used. However, non-verbal representation deserves more focus, to understand how educators should deal efficiently with such representations in their teaching environments.

NON-VERBAL KNOWLEDGE

The Case of Body Knowledge

According to Dewey, mind and body have been perceived by educators as separate entities that may even interfere with each other. According to this notion, bodily activity is considered by educators as an intruder that,

> having nothing, so it is thought, to do with mental activity, it becomes a destruction, an evil to be contended with. For the pupil has a body, and brings it to school along with his mind. And the body is, of necessity, a wellspring of energy; it has to do something. But its activities, not being utilized in occupation with things which yield significant results, have to be frowned upon. They lead the pupil away from the lesson with which his "mind" ought to be occupied; they are sources of mischief. (Dewey, 1916/1966, p. 141)

In a conference I attended, conducted in Birmingham, England, we were taken for a tour of an old coal mine, which has now become a very interesting museum. In the museum which showed how people lived in those days, a small typical class was presented. The objects which drew my attention were the children's chairs and desks. The chairs were connected to the desks so that the children could not change the distance between the chairs and the tables. The tables were split into two parts. In order for the children to sit in their chairs they had to lift one part of the table. After sitting in their seats they let this part of the table back down. Now, it seemed as if they were "locked" in their seats. This structure of the chair and desk well demonstrates the view according to which teachers should suppress their pupils' body during lessons so that the children's bodies do not disturb their minds. Dewey argues that "it would be impossible to state adequately the evil results which have flowed from this dualism of mind and body, much less to exaggerate them" (p. 141). The importance of bodily knowledge is supported currently by cognitive psychology theories where non-verbal mental representations of knowledge in memory, in different modalities, account for how we think and are at least as fundamental as verbal presentation. There is a growing awareness that *bodily knowledge*, which is the kind of knowledge reflected in motor and kinesthetic acts (Reiner and Gilbert, 2000), is "stored" in our body and impacts our behavior. For instance, each one of us has probably experienced not being able to remember a phone number unless we actually dial the number using the phone's key-pad itself. It is as if our fingers 'know' the number better than us. The knowledge of the phone number is somehow embedded in our body. Another example is the knowledge embodied in ball games such as snooker or basketball. Consider for example snooker players. Experienced players know that if they want the ball they hit with their cue stick to stop after clashing with another ball, they should aim their cue stick at the exact centre of the ball. They know that if they want the hitting ball to bounce back, then they should direct the stick to hit the lower part of the ball. They are also aware of the fact that in order to enable the initial ball to continue rolling forward (in the same direction of the ball being clashed) they should direct the stick to the upper part of the ball. In the same manner, a basketball player knows what amount of force and the direction he or she needs to apply to the basketball so that it will enter the basket. These examples demonstrate that body knowledge is a knowledge that we cannot ignore. According to Dewey,

> Before the child goes to school, he learns with his hand, eye, and ear, because they are organs of the process of doing something from which meaning results. The boy flying a kite has to keep his eye on the kite, and has to note the various pressures of the string on his hand. His senses are avenues of knowledge not because external facts are somehow "conveyed" to the brain, but because they are used in doing something with a purpose. (p. 142)

This idea is also supported by Piaget's cognitive development theory (Piaget *et al.*, 1952). One basic assumption underlying Piaget's theory is that the origin of thinking is in sensomotorisch activity (kinesthetic experience) of the physical surroundings. In the process of cognitive development, sensomotorisch activity is assimilated and then appears in the form of mental operations in the stage of concrete operations at the ages of 6–7.

Sometimes we are not aware of our body knowledge (Henry, 1953). For instance, having aquired a particularly high level of skill, an athlete seems to disconnect bodily performance completely from overt cognitive control and the body 'takes over' (Starkes and Allard, 1993). Reiner and Gilbert (2000), referring to the work of Starkes and Allard (1993) argue that it seems as if the body 'knows' something the player 'does not'. Rather than rational propositional knowledge being used, some sort of imagistic, embodied form of knowledge, which is not 'registered' in the conventional manner, is being employed.

There are some studies supporting the impact of body knowledge on the understanding of the learned subject. Helm (1991) studied the effects of learning modalities and found that kinesthetic subjects attained the highest grade averages. Prifster and Laws (1995) described experiments that are possible with kinesthetic devices which help eradicate some of the traditional student misconceptions and provide students with a deeper understanding of basic physics concepts such as Newton's laws. Clement's (1988) findings also support this view in that he showed that embodied intuitions about forces have a role in understanding physics situations. He suggests that knowledge embodied in perceptual motor intuitions is used for physics problem solving by experts.

In her interesting research, *Effects of the Kinesthetic Conflict on Promoting Scientific Reasoning*," based on Piaget's original work, Druyan (1997), tested the effect of kinesthesia on children's learning. Here I will describe experiment 1 of her study concerning the concept of length. Before the intervention, the participants were given two drawings of two paths that had the same starting and ending points. One of the paths was a straight blue path 10 cm long, and next to it was a zigzagged green path 15 cm long. At the start of each path was a picture of a child, and at the end of each path was candy. The subjects were told that the children in the drawing want to reach the candy at the end of their path. Each child claimed that his path was the shorter one, and a third child, standing on the side claimed that the two paths were equally long. The children were asked to decide which one of the children was correct. Was one path shorter than the other or were they equal in length? Those of the children who did not answer correctly on this task were randomly divided into the following three groups:

1. Walking training — the children were asked to walk on a zigzag path (15 m) and a straight (10 m) path.
2. Jumping training — the children were asked to jump on both legs along each of the paths (as in 1).
3. Measured Walking with Peer — each pair of children was asked to walk heel to toe simultaneously along the paths in a pace that was determined by a drumbeat: one on the straight path and the other on the zigzag path (as in 1 and 2), after which they changed places.

After each task the children were asked which of the two, the straight or the zigzag paths, was more difficult.

The posttest included the pretest and two other similar tasks: a straight path with a curved path and a straight path with a broken path. The findings of the experiment suggest that the jumping and the kinesthetic measuring tasks (and not the walking — which is an effortless task) were efficient in promoting the concept of length. According to the author,

> Transferring the change in perception of the concept of length from kinesthetic to a more formal level of presentation such as paper task and to other different patterns indicates a high level of cognitive change achieved through effort-involved training. . . . The advantage of measured walking over normal walking

supports the attitude which relates the importance of measuring activity in the thought process. . . . activating the body simultaneously with measuring length might create more significant brain connections. (p. 1089)

Educators should be aware of the importance of body knowledge. They should provide the children with appropriate sensomotorisch experiences which can then be used as an established basis upon which the correct scientific concepts may be built. For instance, as in Druyan's work, exposing children to following efficient kinesthetic experiences like jumping training or measured walking with peers might help children gain a better understanding of the concept of length. And as Duryan puts it, "To improve science teaching, teachers are encouraged to be more creative in developing and using active strategies for learning" (p. 1089).

Use of Visual Representations

I started the section on non-verbal representation with the concept of body knowledge and argued that a teacher who is aware of the impact that body knowledge may have on the construction of concepts, should design his or her lessons accordingly to take full advantage of kinesthetic experiences to promote the learning process. Body knowledge is one kind of non-verbal representation of knowledge. *Visual representations* may also impact the learning processes. To gain a good understanding of visual representations and learning processes, one must understand the difference between external and internal visual representation as well as the relationships between them. Visual representations of every day life include writing, pictures, and diagrams. A visual *mental* representation, or *imagery* is defined as the "mental invention or reaction of an experience that in at least some respects resembles the experience of actually perceiving an object or an event (Finke, 1989). According to Thomas (1999) imagery is a quasi-perceptual experience: experience that significantly resembles perceptual experience (in any sense mode), but which occurs in the absence of appropriate external stimuli for the relevant perception. Imagery is the process by which humans represent knowledge in their minds. According to Kosslyn (1994), imagery is a basic form of cognition and plays a central role in many human activities ranging from navigation to memory to creative problem solving. The following are classes of imagery abilities (Kosslyn, 1994):

1. Image generation and maintenance — there are three ways in which visual images are created. First, one can recall a previously seen object or event. Second, one can combine objects in novel ways. Finally, one can visualize novel patterns that are not based on rearranging familiar components; one can "mentally draw" patterns that he/she has never actually seen. Once the image is created it can also be retained in one's working memory.
2. Image inspection — people can scan an image and 'zoom in and out' to see different parts of that object.
3. Image transformation — The classic research of Shepard, Cooper, and their colleagues (Shepard and Cooper, 1982; Shepard and Metzler, 1971) demonstrated that not only do mental images exist but also that there are mental operations that

can transform them in various ways. People can mentally rotate objects in images. It was found that people can also "mentally fold" objects in images and otherwise transform them (Shepard and Feng, 1972). In addition, it appears that imagery and perceptual tasks in the same mode would often mutually interfere with one another (Brooks, 1968; Segal, 1971).

This indicates that "visual images have all the attributes of actual objects in the world — that is, they take up some form of mental space in the same way that physical objects take up physical space in the world, and that these objects are mentally moved or rotated in the same way that objects in the world are manipulated" (Eysenck and Keane, 1995, p. 215).

Mental representations are generated from our memories of past perceptual experience, and they develop and change as a result of interaction with present sensory input (Kosslyn, 1994). Thus, what we are able to represent in our memory system is often limited by what we have perceived. Stated differently, there is a connection between a human's ability to construct internal visual representations and the external visual representations to which they were exposed. It is clear that by providing students with efficient external visual representations, educators may help them construct mental visual representations. Such representations may assist them tremendously in dealing with novel problems with which they are confronted. Monaghan and Clement (1999, 2000) found that external visual representations experienced through the use of on-line computer simulations of relative motion, facilitated mental simulation off-line and improved problem solving.

Support for the idea of visual imagery is provided by historians of science who argue that visual imagery played a significant role in many scientific and technological discoveries. Gowan (1978), for instance, states that "in the case of every historic scientific discovery which was reached carefully enough, we find that it was imagery . . . which produced the breakthrough." Einstein himself once wrote, "My particular ability does not lie in mathematical calculation, but rather in visualizing effects, possibilities, and consequences" (Pinker, 1997, p. 285).

Educators should bear in mind that they must be aware of how important it is to encourage children to create efficient visual images that will contribute to their conceptual understanding. This can be done by using rich external representations such as pictures, diagrams, graphs, movies etc. In addition, an educator might also encourage children to use their imaginations to create such visual images. Here is another episode from Feynman's book, describing how Richard's father encouraged him to use his imagination:

> We had the Encyclopedia Britannica at home. When I was a small boy he [the father] used to sit me on his lap and read to me from the encyclopedia. We would be reading, say, about dinosaurs. It would be talking about the Tyrannosaurus Rex, and it would say something like, "This dinosaur is twenty-five feet high and its head is six feet across."
>
> My father would stop reading and say, "Now, let's see what that means. That would mean that if he stood in our front yard, it would be tall enough to put its head through our window up here." (We were on the second floor.) "But his head would be too wide to fit in the window." Everything he read to me he would translate as best he could into some reality.

It was very exciting and very, very interesting to think there were animals of such magnitude — and that they all died out, and that nobody knew why. I wasn't frightened that there would be one coming in my window as a consequence of this. But I learned from my father to translate: everything I read I try to figure out what it really means, what it's really saying. (Feynman, 1988 p. 2)

In the next section, I introduce two visually-based teaching methods: conceptual models and concept maps. Both methods, if used in K-2 appropriately, may contribute to children's construction of meaningful scientific as well as non-scientific concepts.

Conceptual Models

A conceptual model is defined as words and/or diagrams that are intended to help a learner build mental models of the system being studied; a conceptual model highlights the major objects and actions in a system as well as the causal relations among them (Mayer, 1989, p. 43).

According to Mayer, conceptual models:

1. Guide students' selective attention toward the conceptual information in the lesson (i.e. the major objects, states, and actions, and the causal relations among them).
2. Organize the information around coherent explanations (i.e. build internal connections).
3. Integrate the information with existing relevant knowledge (i.e. build external connections).

Students given conceptual-model-based instruction may be more likely to build mental models of systems that they are studying and to use these models to generate creative solutions to transfer problems.

One example is Mayer *et al.*'s (1984, Experiment 1) research. High school students who studied physics were asked to read a 450-word passage on density. Some students were provided with conceptual models whereas others were not. The model showed a diagram of a cube of city air along with a verbal definition of volume and diagrams showing particles in a cube of city air along with a definition of mass. It was found that the students with the model recalled 144% more of the conceptual information, scored 26% lower on verbatim retention tests and solved 45% more of the transfer problems than the control students. In a review article Mayer (1989) argues that since models help students direct their attention toward the conceptual objects, locations and actions described in the lesson, students will improve their conceptual retention. In addition, since models help students reorganize material, they tend to lose the original presentation and will reduce verbatim retention. The most crucial finding about models is that they improve the ability of students to transfer what they have learned to creative solutions of new problems. "The ability to generate novel solutions to new problems is the hallmark of systematic thinking; if students have built models that they can mentally manipulate, they will be better able to solve transfer problems" (p. 59).

For example: while explaining plant growth to a class, a teacher can make excellent use of conceptual models by showing children pictures of a plant in different stages of its growth, with supporting text under each picture. This combination of visual presentation along with text creates a conceptual image like those described by Mayer. For K-2 children, the wording with each image should be as simple as possible. This case would also familiarize the children with the appropriate wording. The teacher may even decide to divide the children into groups and give each of them a different picture, each of a different stage of the growth process. The children can then be asked to put the pictures in the correct order, giving them a chance to participate.

Concept Maps. The concept map, which is a graphical hierarchical representation that links related concepts to form chains of relationships, was developed by Novak in 1977. A concept map contains nodes and labeled lines. Nodes are usually depicted with circles drawn around a term or a concept. The lines between the nodes show which concepts are related. Specific relationships between two concepts are indicated by linking words that are written along the connecting lines. The labeled lines link the concepts to form propositions. These propositions are essential to representing concept/propositional meanings in an explicit hierarchical framework. Novak (1990) argues that concept maps may improve science education in the following four categories: (1) as a learning strategy, (2) as an instructional strategy, (3) as a strategy for planning curriculum, and (4) as a mean of assessing students' understanding of science concepts. This chapter is concerned with the first two categories. First, I will describe the power that concept maps may have on science education. Later, I will argue that using concept maps in their regular forms in K-2 may be problematic. Taking into account the fact that younger children are limited in their literacy skills, I will present a novel way in which concept maps may be used even in kindergarten. I call them *pictorial concept maps*.

The Power of Concept Maps. To understand why a concept map is a useful tool which may tremendously improve teaching and learning, one should first understand how knowledge is mentally represented. It is well known that conceptual knowledge is highly interrelated in nature (Heit, 1997). In addition, people's conceptual structures are widely believed to bear the general properties of hierarchies (e.g. Markman and Callanan, 1983). According to Berlin (1992), hierarchical structures appear to be a universal property of all clusters' categories of the natural world. A hierarchy is a special kind of network in which the only relation allowed between category members is the set inclusion relation. For example, the set of animals includes the set of fish which includes the set of trout which includes the set of rainbow trout (Murphy and Lassaline, 1997). Set inclusion is sometimes called the IS-A relation (Collins and Quillian, 1969): A Mercedes IS-A car and a car IS-A vehicle. The IS-A relation is asymmetric: all cars are vehicles but not all vehicles are cars. In addition, the category relations are *transitive*: all dogs have warm blood, all warm blooded animals are mammals; therefore all dogs are mammals. These properties of hierarchical descriptions enable learning. For instance, if a child learns something about animals in general he or she may now generalize this to the many categories that are under animal in the hierarchy (Murphy and Lassaline, 1997). According to the authors, by being able to place a category into its proper place in the hierarchy, one can learn a considerable amount about the category.

Concept maps indeed are based on the epistemological idea that concepts and concept relationships (i.e. propositions) are the building blocks of knowledge and that internal representations of knowledge are connected in some useful way. The theoretical rationale upon which concept maps are based, according to Novak, are the following two ideas from Ausubel's theory of cognitive learning: (1) new concept meanings are acquired through assimilation into existing concept/propositional frameworks, meaning that new learning occurs through the derivative or correlative *subsumption* of new concept meanings under existing concept/propositional ideas. Indeed, teachers may use concept maps to, "tap into a learner's cognitive structure and to externalize, for both the learner and the teacher to see, what the learner already knows" (Novak, 1984, p. 40). (2) Cognitive structure is organized hierarchically. Constructing concept maps permits one to begin with the most general, most inclusive concept and to show propositional structures in a hierarchical arrangement. In addition, "Learning of concepts is becoming meaningful when we are able to draw relationships between these concepts and other concepts. In fact it is reasonable to assume that the unit of meaningful learning is two concepts plus the linking word(s) that form a proposition, and that the concept meaning grow, differentiate (i.e. become more explicit relatable to more examples), and gain in sophistication as they become embedded in larger and more diverse propositional frameworks" (Novak and Musonda, 1991).

Considering the previously mentioned idea that there is a connection between external and internal visual representation and the unique characteristics of concept maps, it is reasonable to assume that exposing children to concept maps and/or encouraging them to create ones of their own may contribute to their ability to discover connections between concepts and gain a deeper understanding of the subject at hand. In other words, a child who sees the connections between different concepts may also build an efficient coherent internal mental representation of the subject. In addition, I agree that concept maps may also help students to "learn how to learn" and take charge of their own meaning making (Novak, 1985).

According to Symington and Novak (1982), primary-grade children can develop very thoughtful concept maps which they can explain intelligently to others. But what if we wish to present kindergarten and first grade children with the idea of concept maps? At these ages the limited ability of children to read and write may pose a severe barrier to their use. To overcome this barrier I will introduce the idea of pictorial concept maps.

Pictorial Concept Maps. To overcome the existing literacy barrier with kindergarten and first grade children, the concept maps in these cases should involve visual representations of the subjects. Pictorial Concept Maps maintain all the characteristics of the normal concept maps, but they also add graphical representations to the written words. Figure 2a describes a simple concept map of a tree. Adding pictures to the concept map, as shown in figure 2b, transforms it into a pictorial concept map. The addition of pictures is crucial in making the concept maps usable with children of this age. For instance, the graphical presentation of the tree parts helps the children gain a concrete sense of their meanings, along with their names, which are written underneath the pictures. This way, a connection is made between the written term and the part itself. The pictures also enable the child to construct a hierarchal knowledge structure of the subject.

Analogical Reasoning

The idea of visual representation is well connected to analogical thinking. Analogical thinking refers to situations in which people are confronted with problems for which they do not have any directly relevant knowledge. In such cases people may apply knowledge indirectly by making an analogy to the problem. Analogies are usually visualizable and imaginable. Analogical thought involves a mapping of the conceptual structure of one set of ideas (called a base domain) into another set of ideas (called a target domain). In his book *The Society of Mind* Minsky (1985, p. 57) writes,

> How do we ever understand anything? I think by using one or another kind of analogy — that is, representing each new thing as though it resembles something we already know.

I mentioned earlier that Einstein's ability was in visualizing — "certain signs and more or less clear images, which can be 'voluntarily' reproduced and combined" (Hadamard, 1945). He called such thinking "thought experiments." One may recognize analogical thinking in Einstein's thought experiments. For example, he would imagine a man in a falling elevator and then try to see what would happen to the keys in the man's pocket (Rico, 1983, p. 71). This helped Einstein map the conceptual

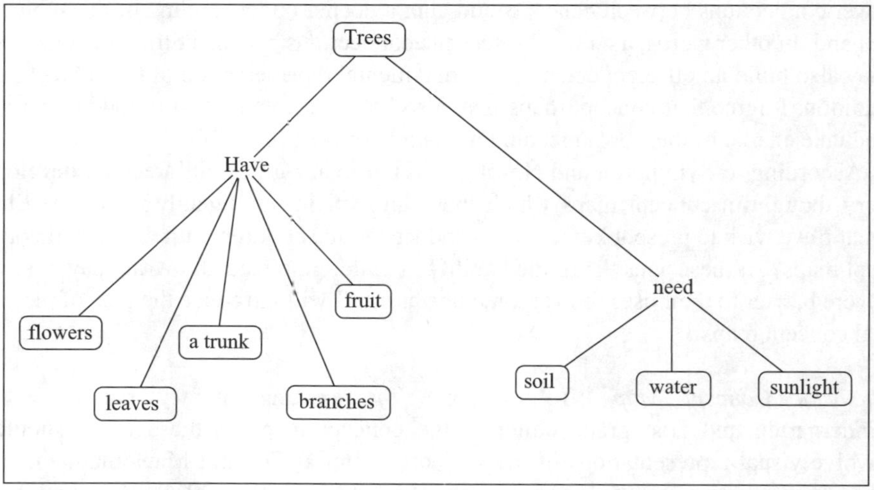

Figure 2a. The tree concept map.

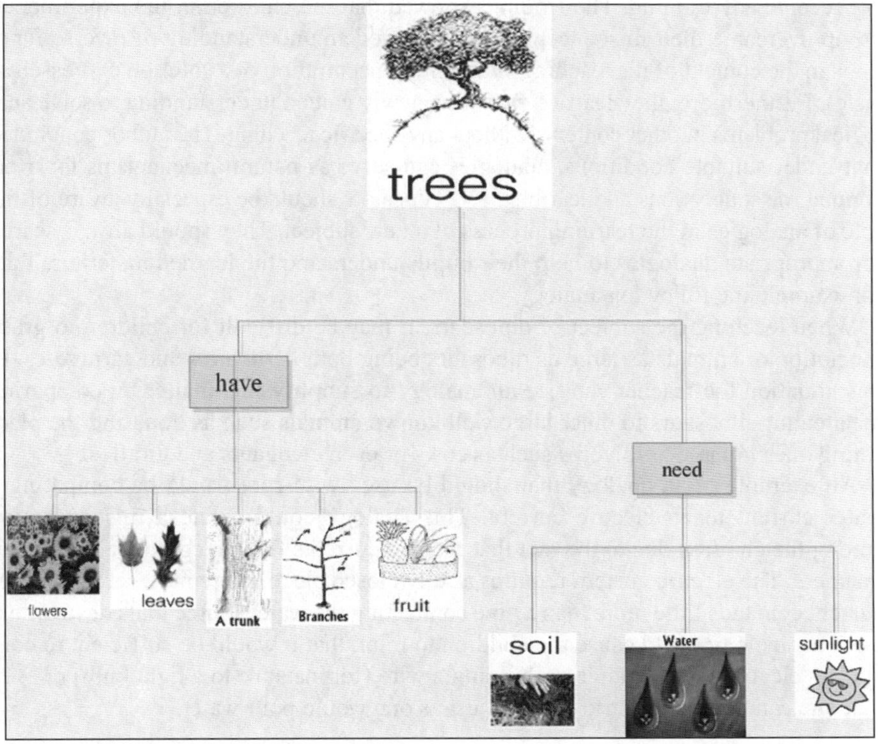

Figure 2b. The tree pictorial concept map.

structure of the keys into his theory of relativity. Another example is the discovery of the benzene structure. Kekule saw the benzene ring as a reverie of snakes biting their tails (Pinker, 1997). He mapped the snake images into the benzene structure. All these analogies are visualized in some sense.

According to Stavy (1991), an effective teaching tool in science which helps correct misconceptions is teaching by analogy. In this way students build on ideas which match their existing intuitive knowledge. In one of the authors' studies (Stavy, 1991, Experiment 1) children from the second, third, and fourth grades were tested individually in the understanding of inverse functions in three contexts: (1) comparing the taste of two sugar-water solutions containing the same amount of sugar but a different amount of water; (2) comparing the temperature of two different amounts of water heated by the same heat source for the same length of time; and (3) comparing the taste of equal-sized bites from two different size pieces of bread, spread with the same amount of chocolate spread. All these tasks represent ratio within different contexts. The research population consisted of the children who could not accomplish all three tasks. Half of that group was treated by being taught the role that the quantity of the water had in the concentration (taste) of a salt water solution; the other half did not receive any training. The findings showed that the children in the experimental group overcame their misconceptions and gained an understanding of inverse functions in the context of the teaching situation (concentration of a solution expressed as taste). Furthermore, they learned from this newly gained understanding to solve analogical problems in other contexts without any specific teaching. The author concluded that under suitable conditions, analogies can serve as natural mechanisms for overcoming misconceptions and learning. K-2 educators should be especially aware of the role of analogies in the learning process of a new subject. They should always search for appropriate analogies to help their pupils understand the learned materials. Take for example the following analogy.

When teaching the subject of dinosaurs, it may be difficult for children to grasp the notion of animals as large as dinosaurs being both herbivores and carnivores. In this situation the teacher can use an analogy to simplify this matter by comparing meat eating dinosaurs to other large well known animals such as lions and the plant eating ones to large herbivores such as cows or even elephants and giraffes.

An example of an analogy that should be used with care would be comparing a water current to an electric current. This analogy could create a misconception among the children due to the fact that the two currents flow in completely different manners. The electric current requires a fully closed electric circuit, whereas a water current demands little more than a pipe connecting the water source to its destination. Such an analogy could cause the children to think that it would be sufficient to connect an electric power source with a single wire (for instance to a light bulb) or even that one can pour an electric current, just as one would pour water.

DISCUSSION

In a thorough review article, Metz (1995) argues that the science curricula at elementary schools emphasizes the "concrete" with a focus on the processes of observation, ordering, and categorization of that which is directly perceivable. Within this approach, abstraction, ideas which are not tied to the concrete and manipulable, as well as planning investigations and determining their results, should in a large part be postponed until higher grades. According to the author, this approach stems from the misinterpretation of Piaget's developmental theory. Close examination of Piaget's work, Metz argues, fails to support this assumption. Elementary school children are capable of grasping some abstract ideas. "They can engage in scientific inquiry and infer new knowledge on the basis of their experimentation. Thus, it is not necessary to emphasize the process of observing, ordering, and categorizing the directly perceivable and concrete, while relegating scientific investigation to latter years" (Metz, 1995, p. 120). According to Novak (1990) there is considerable debate in the science education community as to whether or not young children are capable of understanding abstract concepts such as *energy*, *molecules*, or *evolution*. He argues that the results of his early studies suggested that the primary limitation for young children is not their "cognitive operational capacity," as indicated in the work of Piaget (1926), but rather the quantity and quality of their relevant knowledge acquired through experience and instruction. It does not matter whether or not Novak's explanation complies fully with regard to Piaget's work as suggested by Metz. The important notion to be taken from Novak's work is that with carefully designed instructions, six and seven-year-old children can acquire a useful level of understanding of *any* basic scientific concept, including concepts of energy and energy transformation, the particulate nature of matter, and the conservation of matter/energy. This concurs with Jerome Bruner's (1963) claim that any idea could be taught in some intellectually honest form to children of any age.

Taking into account the fact that children are capable of learning science, and especially that it is in the educator's hands to find good and efficient ways to teach science as early as K-2, this chapter helps shed some light on the subject. The manner by which science should be taught to children is not as obvious as might be thought at first glance. Science educators have for decades been struggling with the issue of how to implement John Dewey's (1910) call to teach science to children in a way that emphasizes method over content (Champagne and Klopfer, 1977). In this chapter I have presented some educational approaches that might help educators gain a better sense to the question, *How should science be taught to K-2 children?* In addressing this question I have presented both teaching strategies and theoretical explanations as to why these strategies should be used. I began with stressing the importance of the development of investigation skills. By doing so, I hope to represent the West's approach to teaching according to which,

> In the West we generally encourage children to try to solve problems and to contrive objects on their own. We see it a positive development when a child sports a pair of adult glasses or monkeys around with a key that is destined for a specific slot. Westerners have gained a certain hegemony in the contemporary world by exploring, trying out new approaches, experimenting and revising — whether in pursuing science and technology, or in exploring the ocean and outer spaces. (Gardner, 1999, pp. 94–95)

The idea that investigative skills may be implemented through *problem-centered* techniques such as problem-based learning, and learning through projects is explored. Problem centered learning is also congruent with *constructivisim*, which asserts, among other things, the importance that prior knowledge has in learning. Such learning environments encourage students to elaborate on their own knowledge and invite students to negotiate meaning in small group situations and then negotiate a consensus in the whole class setting (Wheatly, 1991). This also well expresses the view of those who have used constructivism according to which knowledge is personally constructed but socially mediated (Tobin and Tippins, 1993). As was mentioned in the problem-based teaching section, from Holyoak's argument in particular, it is reasonable to assume that the children who are used to being confronted with problems definitely acquire general cognitive abilities that help them deal with problems in a wide variety of domains. Introducing problems alone, however, is not enough. Scaffolding is a necessary process that helps the child build cognitive abilities. Nevertheless, scaffolding can be a vague term for an educator. In this chapter I described some scaffolding strategies. Although these strategies may help children develop problem solving skills, something else might be needed in order to address Mayer's concern. The author begins his article *Models for Understanding* with the following citation from Luchins and Luchins (1970, p. 1):

> Why is it that some people, when they are faced with problems, get clever ideas, make inventions and discoveries? What happens? What are the processes that lead people to such solutions? What can be done to help people to be creative when they are faced with problems?

The author articulates that one promising technique for helping students learn new material in approaches that allow them to be creative with problems is the use of conceptual models. This was one reason why in addition to problem-centered strategies, I chose to elaborate on the idea of none-verbal representations, of which Mayer's concept models are only a small part. I also further discuss the idea of concept maps and kinesthetics, which I believe may contribute tremendously to inculcating problem solving skills as well as to the understanding of basic scientific concepts. Comprehending these approaches might help educators enrich their scientific *pedagogical content knowledge* (PCK) (Shulman, 1987). PCK "represents the blending of content and pedagogy into an understanding of how particular topics, problems, or issues are organized, represented, and adapted to the diverse interests and abilities of learners, and presented for instruction"(p. 8).

I used Feynman's story to illustrate some of the issues explored above, not because I believe that anyone that receives the same learning experience as Richard will get a Nobel Prize, nor is it because I believe everyone should be a scientist. It is my belief that by teaching by these means, we might be able to help children to exploit their full cognitive potential in whatever field they choose. The story shows that these approaches can be implemented. If Richard's father could do it, then an educator, who is exposed to novel educational ideas, should have absolutely no problem doing so as well.

Another example of how Feynman's father taught him science will also be used in the next chapter, which introduces the idea of using technological apparatus to teach science.

CHAPTER 3

WHEN LEARNING SCIENCE BY DOING MEETS DESIGN AND TECHNOLOGY

LEARNING BY DOING

I hear, I forget. I see, I remember. I do, I understand. (a Chinese proverb)

According to Schank (1996) there is only one way to learn how to do something and that is simply to do it. If you want to learn to play checkers, solve a mathematical problem, prepare a pizza, drive a car, or design a building you must have a go at doing it. Humans are natural learners. They learn from everything they do. This is probably what Dewey had in mind when he wrote,

> Thinking is the accurate and deliberate instituting of connections between what is done and its consequences. . . . The stimulus to thinking is found when we wish to determine the significance of some act, performed or to be performed. (Dewey, 1966/1916, p. 151)

The notion of learning by doing somehow challenges the old philosophical belief that humans are rational beings and that the laws of logic are the laws of thought. According to this view, if we humans are rational, it would be enough for us to learn abstract concepts and rules in order to apply them to a variety of situations which we encounter in everyday life. Also, *doing*, along with when and where we experience a situation where rules or concepts apply would have little, if any impact on the learning process. In other words, knowing the concepts and rules, which contain small pieces of knowledge and thus allow economy of storage, could be enough for dealing with all of the situations where the learned concepts and rules may be used. This is exactly what *rule-based reasoning* is. However, it has been found that people have difficulty applying concepts and rules to particular situations. One reason is that concepts as well as rules are expressed too abstractly and may be unintelligible. It is the *doing* in a context which makes the concepts and the rules we learn meaningful to us. Learning by doing finds support also in the *case-based reasoning* theory. According to case-based reasoning, reasoning is a process of retrieving examples rather than applying rules. In terms of case-based reasoning, by *doing* we acquire experience, or more specifically — cases which, as opposed to rules, contain large chunks of knowledge which are tied to a context.

> Experiences, or cases, are a critical element in understanding what is learned when one learns by doing . . . a learner is interested in acquiring sufficient cases such that he can learn to detect nuances. He wants to be in a position to compare and contrast various experiences. To do this, he needs to have had those experiences, and he needs to have properly labeled those experiences. The labeling process is what we refer to as indexing. (Schank, 1996)

The more cases we acquire, the index we will construct will be better, richer, and more efficient. This will eventually lead to a better remembering of an old case to use for decision-making with a new case.

Lave and Wenger's *situated learning* theory also supports the notion of doing in learning. Here, individual learners do not gain a discrete body of abstract knowledge which they can then transport and reapply in later contexts. Rather, they acquire the skill to perform by actually engaging in the process. Initially, people have to join communities and learn at their peripheries. As they become more competent they advance closer to the 'centre' of that particular community. Thus, according to the situated learning theory, it is irrelevant to talk of knowledge that is decontextualized, abstract or general. The nature of the *situation* impacts significantly on the process. Lave and Wenger illustrate their theory by observing different apprenticeships: Yucatec midwives, Vai and Gola tailors, US Navy quartermasters, meat-cutters, and non-drinking alcoholics in Alcoholics Anonymous. For instance, the Yucatec Mayan midwives learners in Mexico were usually the daughters of experienced midwives, with knowledge/skills being handed down within the families. The learning process was informal and part of daily life.

Schank (1996) argues that since learning by doing is how we naturally learn in real life, motivation is never a problem. We learn because something makes us want to know. What does this all tell us about education? It tells us that when designing a curriculum, we must keep in mind what it is that we are trying to have students who will go through that curriculum be able to do. To put it another way, we need to transform all training and education to make it look, and feel, like doing. However, according to Schank, there has always been a great deal of lip service given to the idea of learning by doing, although not much has been done about it in practice. The author cites John Dewey who, almost a century ago, wrote in his famous book *Democracy and Education*:

> Why is it, in spite of the fact that teaching by pouring in, learning by a passive absorption, are universally condemned, that they are still so intrenched in practice? That education is not an affair of "telling" and being told, but an active and constructive process, is a principle almost as generally violated in practice as conceded in theory. Is not this deplorable situation due to the fact that the doctrine is itself merely told? It is preached; it is lectured; it is written about. But its enactment into practice requires that the school environment be equipped with agencies for doing, with tools and physical materials, to an extent rarely attained. It requires that methods of instruction and administration be modified to allow and to secure direct and continuous occupations with things. (p. 38)

According to Schank (1996) education today has not changed very much from Dewey's days — it is still an affair of telling and being told. School has no natural motivation associated with it. Students go there because they have no choice. He gives two main reasons as to why learning by doing is not our normal form of science education. First, is the lack of "doing devices." The second reason is that educators and psychologists have not really understood why learning by doing works, and are thus hesitant to insist upon it. "They can't say exactly what it is that learning by doing teaches. They suppose that it teaches real life skills, but what about facts, the darlings of the 'drill-them-and-test-them' school of educational thought?" (Schank, 1996, pp. 295–296).

I do not fully agree with Schank that students go to school only because they have no choice. I believe that most children do find school to be a place where they enjoy.

They do learn a lot and school definitely plays an important role in their cognitive as well as emotional development. I also disagree with Schank that teaching in school today is an affair of telling and being told. On the contrary, huge efforts are being invested to respond to the call of national and international reports such as the American Association for the Advancement of Science (AAAS) (1993) according to which, if the next generation is to become scientifically literate, then learners need to become actively involved in exploring nature in ways that bear a resemblance to how scientists themselves do their work. This indeed concurs with Wolpert's comment that

Science is a special way of knowing and investigating and the only way of appreciating the process is to do it. (Wolpert, 1997, p. 21)

The problem is not that schools do not encourage *doing*. Rather, as I shall show, the problem is how learning by doing is *implemented*. The following citation clarifies this argument,

Yet, although elementary and middle schools are increasingly exposing inquiry-based or "hands-on" science, the objective of authentic experimentation is rarely pursued in school. Instead of extended and systematic work to explore a personally meaningful phenomenon or question, students in hands-on programs too often engage in a string of unrelated, one-period, 40-min . . . activities that emphasize the use of materials and equipment but are often poorly or entirely unmotivated from the student's point of view. Although there may be an overall design or plan behind the sequence, it is typically motivated by the structure of the scientific discipline. Because students do not share this understanding of the overall structure of the discipline, the logic behind the sequence may be apparent to teachers but a mystery to students. (Schauble *et al.*, 1995, pp. 132–133)

The authors argue that even a hands-on activity that occurs in a laboratory setting may be introduced to students as exercises rather than experimentations their emphasis on drill and mastery, practicing disembodied skills and the conduct of procedures with meanings which are not clear to the participants.

According to Moscovici (1998) the explanation for this situation stems from the teachers' lack of abilities. He reported that the general perception expressed by prospective elementary school teachers in his research was that they couldn't use techniques consistent with inquiry, as they were never involved as students in such processes. They also feared that their perceived weak background in science did not support such techniques. If they were going to teach science, they felt more comfortable with a series of disconnected activities, or what he called "*activity mania*."

There appears to be some confusion among three key components: learning, doing, and learning by doing. Schools may provide learning environments that do not encourage doing. In other cases, which I believe is the most common problem, schools may offer hands-on activities to their students — this is doing. This way of doing, however, is not always efficient in leading children to meaningful learning. Doing in such cases is detached from meaningful learning. Doing may contribute tremendously to learning. But, it should be taken into consideration that educators need to design efficient doing activities that will fit children's needs and indeed contribute to their learning. Fig. 1a and 1b demonstrate this situation. Fig. 1a illustrates the situation where there is doing, but it may not necessarily lead to meaningful

Figure 1a Doing Separated From Learning:
Doing without meaningful learning

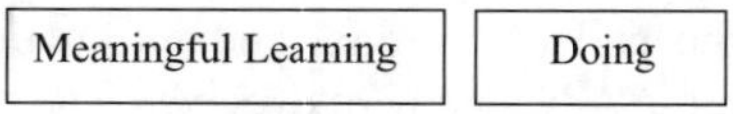

Figure 1b Doing Leads to
Meaningful Learning: Learning by Doing

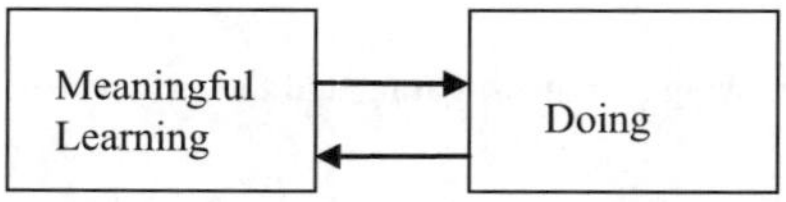

Figure 1. Relationships between doing and meaningful learning.

learning. Of course, there is always some learning achieved when doing. In such cases the potential of learning is not fully exploited. In Fig. 1b there are two arrows, one from doing to learning, and another which goes the other way around, from learning to doing. This demonstrates that in an efficient learning environment, doing may lead to meaningful learning and, in turn, we learn more and as a result can do more. I agree with Haigh *et al.* (2005), who state, "doing science has been a central theme in much international science education literature and, while there appears to be some consensus on the doing, there is less on the what for" (p. 215).

So far I have described the need to implement the *learning by doing* approach in science education. I also warned against detaching *doing* from *meaningful learning*. In other words, I argued that doing by itself should not be our aim but should rather serve learning in ways to make it meaningful. There are many ways which one can implement the *learning by doing* approach. In this chapter I will thoroughly discuss the learning of science via technology, especially through designing, building, and evaluating simple mechanical devices. It is my view, as I hope to convince the reader, that such an approach, if implemented appropriately, well fit the teaching of science both in kindergarten and primary schools. First, I shall first explain the terms *technology*, and *design*. I shall then show how one can use technology and design to enhance the learning of science.

THE TERM TECHNOLOGY

In his excellent book, *Teaching About Technology — An Introduction to the Philosophy of Technology for Non-Philosophers*, de Vries (2005) writes "I have abstained from any effort to give a definition of technology. For those who are looking for a definition there are thousands out there to choose from and I do not think I can come up with the one that beats them all" (p. 11). To gain a sense of what technology means, it is worthy to look at the term's origins. *'Technology'* derives from two Greek words: *'technē'* and *'logos'*. 'Techne' means art, skill, or craft. Specifically, a 'technē' is a skill or art that is learned, a professional competence rather than a natural talent. This means that 'technē' involves the practical skills of knowing and doing. The root 'logos' means 'word', but, particularly, a word that comes from rational thought (the Latin translation of 'logos' is 'ratio' from which we derive not only 'ratio' but also 'rational'). Thus, 'logos' can also mean speech, an account, or a discourse, as well as reason in itself. Technology, thus, encompasses reasoned application (Herschbach, 1995). Although the origin of the term technology relates to both knowledge and doing, the term "technology" in the English language, which acquired limited usage in the late 19th century, referred in those days mainly to applying science to making and using artifacts. Today, however, there is increasing emphasis on the importance of knowledge in defining technology (Layton, 1974; McDonald, 1983). de Vries (2005) takes the term "technology" in the broad sense as "human activity that transforms the natural environment to make it fit better with human needs, thereby using various kinds of information and knowledge, various kinds of natural (materials, energy) and cultural resources (money, social relationships, etc.)" (p. 11). To understand more fully the meaning of technology one should understand the relationships between science and technology.

Views Concerning the Relationships Between Science and Technology

Fensham and Gardner (1994) identified the following four possible propositions about the relationship between technology and science. The first proposition, in my opinion, considers and emphasizes mainly the practical aspect of the term technology, i.e. the *'technē'* by neglecting the knowledge component, i.e. the 'logos.' The second proposition also takes the 'logos' aspect of the term technology into account. The third and forth propositions consider both the *'technē'* and the 'logos' aspects of the term technology.

1. Science has historical and ontological priority over technology — in this view, scientific knowledge is necessary for technological capability and is acquired first. There is ample evidence for this claim. For instance, the electric industry in the 19th century and the nuclear power industry in the twentieth obviously rest on strong scientific bases. This view is well expressed in Feibleman's (1972) distinction between pure science, which uses the experimental method in order to formulate theoretical constructs, explicate natural laws, and expand knowledge; applied science which focuses on applications for purposeful activity; and technology which puts applied scientific knowledge to work.
2. Technology has historical and ontological priority over science — in this view, technological knowledge is necessary for developing scientific knowledge. There is some evidence for this claim. For instance, cannon balls launched from catapults were rounded in order to improve accuracy centuries before the physical principles of projectile motion and air resistance were formulated; Chinese built firework rockets in advance of any established theory of rocket propulsion, steel was made prior to the full understanding of the metallurgical process; and Bell's

telephone system, which was dependent on the electrical properties of carbon which were unknown to science at the time he used it. Moreover, medieval developments in clock-making laid the foundation for our modern concept of time. According to Cajas (1999) engineers use science for their specific needs. Their 'use' of science is not the simple application of universal knowledge to particular problems. Rather, they *construct* knowledge for specific situations illuminated by practical and mundane information. Further, Mitcham (1994) argues that the idea of a machine, the concept of a switch, invention, efficiency, optimization; the theories of hydraulics and aerodynamics, of kinematics and cybernetics, of queuing, information, and network theory — are all inherently technological. Such ideas are not found, according to the author, in scientific fields, but rather in technological ones. The author reaches the following rather provocative conclusion, "Indeed, the use of mechanics in science (as in Newton's "celestial mechanics") can reasonably be argued to be derived from early modern technologies (of, especially, clocks), so that science in some senses might be described as applied technology" (p. 96).

3. Technology and science are independent systems of thought and practice — Drucker (1961, in Fensham and Gardner (1994)), who proposes this view, for instance, argues that until modern times, science and technology were independent. History shows that there were cases in which artifacts and procedures co-existed with incompatible scientific beliefs. Then if an innovation was vaguely incompatible with a scientific theory, this was not necessarily disturbing. Although eyeglasses had been in use since the late thirteenth century, Galen's theory of vision, which ruled out any possibility of correcting visual defects, continued to be taught for three centuries.
4. Technology and science engage in two-way interaction — according to this interactionistic view, technologists and scientists learn from each other. This is done either over a long period of time, or contemporaneously through shared knowledge gained through social networks, or through working in close proximity on a common task. Indeed, in modern fields such as electronics, radio astronomy, computing and genetic engineering, scientists and technologists do in fact work together.

According to Roth (2001) science and technology are deeply related domains, part of a (semiotically) seamless web that integrates any distinction. To clarify this notion Roth claims that "gains in the theoretical knowledge about the telescope evolved together with gains in the understanding of its mechanical properties. Thus, Kepler contributed to the further development of the telescope by designing new types and by formulating the law of the inverse relationship between light intensity and square distance" (Roth, 2001, p. 770).

It appears that today, technology is conceived as more than artifact and or a series of techniques and processes. Technological knowledge is indeed considered to have its own abstract concepts, theories and rules, and its own structure and dynamics of change. However, one should bear in mind that (1) technological knowledge is essentially applicable to real situations and that; (2) the defining characteristic of technological knowledge is its relationship to *activity*. Technological knowledge arises from, and is embedded in, human activity. As Landies (1980) observes, while the intellect is at the heart of the technological process, the process itself consists of "the acquisition and application of a corpus of knowledge concerning technique, that is, way of doing things" (p. 11). "It is through activity that technological knowledge is defined; it is activity which establishes and orders the framework within which technological knowledge is generated and used" (Herschbach, 1995).

Surprisingly, although technology is connected to human activity, the education of technology related aspects are not always connected with activity. In many educational curricula which try to show the connections between science and technology, students are exposed to a technological system, however they are not at all required "to *do*."

In cases where *technology is learned scientifically*, we are actually missing out on a significant opportunity to learn by *doing*. It might be that the term design, which entered the scene of technology education, e.g., the Design and Technology curriculum which will be described later, emphasizes and highlights the doing aspect of technology. This is ironically the case, since the term design was originally meant to emphasize that technology is not merely a technical subject but rather a subject which requires higher order thinking, as is the case with design.

THE TERM DESIGN

The Oxford dictionary defines *design* as a mental plan. A plan or scheme conceived in the mind and intended for subsequent execution; the preliminary conception of an idea that is to be carried into effect by action; a project. From this definition one may understand that designing is reified intentional activity. This idea is well expressed by de Vries (2005) concept of *design plan* which he describes as follows,

> A designer has the *intention* of realizing a certain new artifact that can fulfill a certain function. The designer has *beliefs* about the physical properties of such an artifact and how they could make the artifact fulfill that function. Then the designer sets up a *sequence of actions*, a plan, of which (she) believes that it will result in the artifact. The designer has the *disposition to act* accordingly, and when no other considerations show up, (s)he will act accordingly. (p. 60)

The capacity for design is analogous to the capacity for language. Design ability, like language ability, reflects a capacity that everyone possesses at least to some degree, definitely not, the possession of a gifted few (Roberts, 1994).

> We all, as instances, try to create an environment which reflects our aspirations; use tools and materials purposefully; make judgments about which objects and places we like or dislike; find ourselves moved and excited by fine things that other people have made; respond to the visual messages of advertising, products, signs, buildings, films, television; and create visual images by photography and make qualitative judgments about which ones are 'successful' or which ones are 'unsuccessful'. (Roberts, p. 173)

Mental models are the 'language' of design. They contain knowledge which may be represented by propositions as well as knowledge such as sketches, drawings, and diagrams. The latter kind of knowledge, the non-propositional one, contains a richness that could never be entirely expressed in propositions (de Vries, 2005). This means that designing requires one to form mental images in his or her mind.

Mistakenly, design is often identified with one of its languages — drawings. It is important to bear in mind that design is done essentially in the mind, and making drawings or writing notes is a recording process (Report on Engineering Design, 1961). External visual representations such as drawings, diagrams, mock-ups, and prototypes may help in the design process as well as in expressing the internal process. According to Mitcham (1994) the mental effort required in the designing process is something distinct from knowing or coming to know in a scientific or theoretical (or even technological) sense, because it does not terminate in an interior cognitive act. "Designing ends with Aha! Let's make it this way. Let's go with this design" (Mitcham, 1994, p. 221).

Crismond (2001), based on the literature, argues that design, like scientific inquiry, engages the core strategies of analysis, synthesis, and evaluation, which appear as the three highest-order educational objectives in Bloom's taxonomy (1956). Acknowledging the potential that children who are exposed to design activities are likely to develop higher order thinking skills, was probably one of the reasons which led the American National Research Council to create "Science and Technology" content standards for its National Science Educational Standards, which advocates that "As a result of activities in grades K-4, all students should develop abilities of technological design" (p. 135). The report continues by saying that "This standard helps establish design as the technological parallel to inquiry in science. Like the science as inquiry standard, this standard begins the understanding of the design process, as well as the ability to solve simple design problems" (p. 135).

I started this chapter by explaining the rationale of *learning by doing*. I then suggested that one way to implement learning of science by doing is through starting the learning process by engaging students with simple mechanical artifacts. In the next section, I will argue that by neglecting to expose children to design and technology activities within science courses, educators miss a fine opportunity to teach science effectively.

APPROACHES TO TECHNOLOGY EDUCATION

In a review article, Technology education in Western Europe, de Vries (1994) summarizes eight approaches to science education. Four of these approaches that relate to elementary science education, are presented here: (1) The Craft-Oriented Approach, (2) the Design Approach, (3) the Science Technology Society (STS) Approach, and (4) the Applied Science Approach.

The Craft-Oriented Approach

Central to this is making things. Children are given work drawings in which the design has been elaborated in detail, including the materials and procedures. Most of the time is spent making work pieces. The concept of technology developed by this approach is an instrumental one: technology is a way of making things. Design does not play a role in this approach. It emphasizes the *doing* aspect of technology.

Kindergarten and primary school children usually have a good deal of experiences with this approach. For example, they build dinosaurs, cars, boats, and war hero structures using pre-designed Lego kits. In addition, children build artifacts such as wooden brick structures, dolls from cloth pieces, or cardboard cars. However, these activities are artistic in nature. They do not include all of the previously mentioned stages of design. It is my view that such activities are very important. However, as this chapter portrays, it is my opinion that children should also be exposed to some scientific concepts relevant to the structures as well as to more systematic design activities.

The Design Approach

This approach is usually an extension of the craft-oriented approach. In addition to craft-oriented activities, designing skills are also implemented. The children are provided with design problems which they have to solve in a more or less independent manner.

The Design and Technology (D&T) curriculum exemplifies this approach. It was developed in a national movement in England and Wales during the 1980s, and in 1990 the United Kingdoms' National Standards (DESQWO) added "Design & Technology" as a required subject for all students (Department of education, 1990). This approach aims to make students responsible for major decisions about: what kind of artifact or system is needed, what the product will look like, how it will work, and how it should be produced. D&T offers the potential for children to construct, apply, debate, and evaluate models, rather than simply to absorb transmitted information about them. When students engage designing, they have both the opportunity and reason to engage in cycles of model construction and revision (Lesh *et al.*, 1992). In D&T activities, students typically execute the following stages:

1. **Identifying** a need or a **problem** to be solved;
2. **Selecting** an optimal **solution**;
3. **Constructing** a **prototype**;
4. **Testing** and **redesigning**;
5. **Manufacturing** and finally;
6. **Evaluating** (Layton, 1994, See Example 1, for instance, pp. 9–10).

The Investigating and Redesigning (I&R) Approach

Recently, an interesting approach to design and technology was developed by Crismond (2001) — *The Investigating and Redesigning (I&R) approach*. It aims at offering a bridge to help students reach the steps of D&T described above. According to the author, design tasks are often frustrating for novice designers. A sequence of Investigating and Redesigning (I&R) aims at helping less experienced students avoid the feeling of frustration and futility often encountered when first doing design (Schon, 1987). According to Crismond (2001) I&R provides a scaffold via case-based reasoning (Kolodner, 1993) by giving subjects multiple exemplars of working products to investigate and analyze before redesigning. Crismond (2001) argues that with working devices in hand, naïve designers identify features and machines that can be copied or adapted. Using these methods, students can focus their attention on the overall design approach, the scientific concepts and principles embedded in the redesign challenge, rather than on the design task which is still difficult for them. In I&R activities, students are engaged in the following steps:

1. 'Messing about' with the products: **identifying** novel devices, **clustering and ranking** devices, **using** devices and **learning** about them,
2. Explaining the mode of function of these devices: **analyzing** how products work
3. Designing experiments: **listing** the features of an ideal device, **planning** a product comparison,
4. Redesigning devices: **redesigning** the device and **reflecting** on it (Crismond, 2001).

This method is particularly important for K-2 children who are definitely considered novices. A teacher may use such an approach in addition to the acceptable D&T activities. Examples are provided at the end of this chapter. The design approach emphasizes both aspects of technology — doing and logos.

The Science Technology Society (STS) Approach

This approach is an extension of the applied science approach, but pays more attention to the human and social aspects of technology. In this approach students learn that not only does technology influence both science and society, but is also influenced by them. It presents human/social and scientific aspects of technology. However, design does not always play an important role. The user's perspective is the usual approach to understanding technology (Gardner, 1992, in de Vries, (1994)). This means that the *doing* aspect of technology, which is the essence of technology, is hence ignored. Therefore, it is my opinion, that the term technology in the title, *Science Technology Society*, is misleading.

The Applied Science Approach

In this approach, the learning of scientific phenomena starts with asking questions about a certain product's functioning. This approach was developed by science educators who looked for ways to make science more relevant to students. They believed that those questions about the product's function would motivate students to learn scientific topics. However, practical work is regarded, in this approach, as less important than the cognitive elements of education. Creativity and design are almost absent. In addition, the concept which is emphasized is that technology depends strongly on science. Again, the *doing* aspect is ignored.

In the countries where this approach is executed, both the craft oriented approach and the design approach belong solely to technology education. This means that there are two different subjects in the school curriculum: the sciences — biology, chemistry and physics, and the technology. Both are subjects taught separately in the curriculum. The STS and the applied science approaches belong solely to scientific subjects. However, in such curricula, the students do not design or build technological artifacts, but rather 'talk about them.' This contradicts the seamless web approach according to which science and technology are deeply related domains.

Even though science includes some technological aspects and vice versa, no relation to science is portrayed in the design and technology curriculum — both are presented as separate subjects to the child. By doing so, it is my opinion that we educators, are making a mistake! We cannot on the one hand, write in academic journals that science and technology are part of a seamless web that integrates any distinction, and on the other hand teach science and the technology as two completely separate subjects, with separate teachers, separate grades, etc. In what follows I shall introduce my view that one of the ways that we should teach science is by engaging children with simple artifacts (designing and building) at the beginning of the learning process.

SCIENCE EDUCATION VIA TECHNOLOGY: A NOVEL APPROACH TO SCIENCE TEACHING

If science and technology are indeed part of a seamless web, it is reasonable to believe that technology may be learnt only after children have gained some scientific background. However, it is also reasonable to assume that the opposite is also valid. This means that one may start the learning of science from gaining some technological knowledge first. Indeed, recently, the question of whether technology-centered activities afford a learning environment that scaffolds students' learning of science is gaining increased attention among educational researchers (e.g. Layton, 1994; Roth, 2001). Although technology is not a new player on the educational scene, the idea of teaching scientific concepts through technology is quite new. How many times has the reader seen children design, build, evaluate and redesign artifacts at the beginning of the learning process of a scientific concept within the science class? The current chapter is dedicated to the advancement of the *technology-first* approach. It is not my belief that this is the only way to teach science, but rather, that this is an efficient strategy, which educators unfortunately do not utilize in the science class. Moreover, this leads me to suggest, and I will return to this point further on in the discussion section, that we educators need to rethink how we teach science design and technology, and move towards one course — Science Design and Technology.

To get an insight of what the advantages of technology based science teaching might be, let me refer to one of Richard Feynman's stories in his book *"What Do You Care What Other People Think?" Further Adventures of Curious Character*. The story describes how one day, as a little child, he was playing with an "express wagon," a little wagon with a railing around it. Richard found the behavior of a ball inside the wagon to be rather interesting and went to ask his father.

> Say, Pop, I noticed something. When I pull the wagon, the ball rolls to the back of the wagon. And when I'm pulling it along and I suddenly stop, the ball rolls to the front of the wagon. Why is that?. (p. 5)

For the purpose of the current chapter I consider the wagon with the ball inside as a kind technology system that Richard investigated. The technology system caused

Richard to think of a phenomenon which appeared to be related to one of the most fundamental principles of physics. Moreover, the technology system enabled communication between Richard and his father. Both could relate to the same concrete phenomenon — the ball inside the wagon. The father even described a kind of thought experiment,

> If you look from the side, you'll see that it's the back of the wagon that you're pulling against the ball, and the ball stands still .... It doesn't move back. (p. 5)

In addition, Richard could immediately check his father's explanation,

> I ran back to the little wagon and set the ball up again and pulled the wagon. Looking sideways, I saw that indeed he was right. Relative to the sidewalk, it moved forward a little bit. (p. 5)

The wagon with the ball inside, indeed enabled to shuttle between the concrete (the wagon and the ball) and the abstract, as Richard's father taught him the inertia principle. This example, I believe serves as a good demonstration that Richard's learning process of the inertia principle began with dealing with a technology system. One can take this idea even one step further and think of involving the children with the designing and building of a technological system before they learn the scientific principles involved.

In a research aimed at investigating successful *science* activities, Appleton (2002) found that although many primary school teachers were reluctant to teach science — partly due to their lack of confidence and background in science knowledge — a significant number went on to explain how teaching science using "activities that work" enables them to actually teach it with some confidence. The following is a description of an "activity that works", made by Rhonda, a sixth grade teacher:

> [In] year six [the] focus is on energy, and so for one of the [activities] for the electrical energy section, they designed a car or some sort of model to work with electricity. And I extended it and they had to have a switch, which they had to make — they couldn't use a bought switch. They had to present a report on [the car project] .... And it really worked well, because it wasn't directed from me in any way. All they were told was, "this is what you have to have in it and design some sort of model." (p. 397)

Based on such declarations, Appleton (2002) had concluded, that although defined as "science" activities, "activities that work" have rather technological characteristics: they are hands-on, have a clear outcome or result, encourage manipulation — in order to achieve a "right" outcome, and finally — activities that work lend themselves to integration. The author argues, that "activities that work" may be a substitute or supplement to science pedagogical-content knowledge for primary school teachers, who lack other resources for attainment of such knowledge. In the next section, I shall present eight reasons as to how starting from technology is efficient when science concepts are taught.

Reasons for Technology-Based Science Teaching

1. Children tend to employ engineering models of inquiry rather than scientific models. Schauble *et al.* (1991) distinguished between two kinds of experimentation that children use when conducting scientific experiments: engineering and scientific. It is my opinion that the idea may be referred not only to experiments but also broadened to the "inquiry". So, it is my view that children utilize two models of inquiry: Engineering and scientific.

a) ***Engineering Model of Inquiry***. The child's goal in such an inquiry is to produce a desirable outcome. For this purpose the child manipulates and optimizes mostly those variables which he or she believes might impact the result and contribute to achieve the best outcome. Usually, in this type of inquiry, the inquiry reaches an end and is terminated when an outcome is achieved that meets some criterion for acceptability.
b) ***Scientific Model of Inquiry***. The child's aim in this type of inquiry is to understand the role of each variable in order to understand the relations among causes and effects. For this purpose, before reaching a conclusion, the children choose a procedure that exhaustively evaluates all of the involved variables — including those variables that they do not believe play a causal role. The inquiry process terminates in such a model only after the child has completed a systematic set of tests for every variable that could play a role in the system being investigated.

According to Schauble *et al.* (1991) " "Engineering" of this kind arguably has wider applicability to everyday purposes, and may thus be developmentally prior to the more analytic form of thinking involved in scientific inquiry" (p. 860). This might explain Appleton's (2002) conclusion, discussed previously, according to which scientific activities that work have technological characteristics. Schauble *et al.* find support for this idea in Dewey's (1913), which distinguishes between two kinds of scientific activities: practical exploration for the purpose of achieving a desired effect, and investigation for the purpose of achieving scientific understanding:

> It is commonplace that the fundamental principle of science is connected with the relation of cause and effect. Interest in this relation begins on the practical side. Some effect is aimed at, is desired and worked for, and attention is given to the conditions for producing it. At first the interest is bound up with a thoughtful effort, interest in the end or effect is of necessity transferred to interest in the means — the causes — which bring it about. (p. 83)

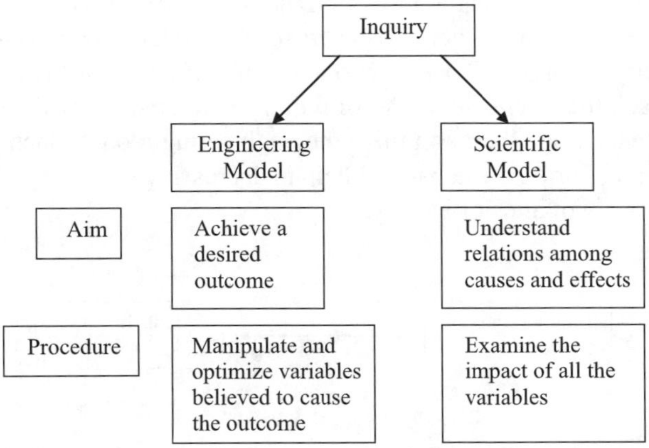

Figure 2. Differences between engineering and scientific models of inquiry.

Here are several examples taken from the literature that illustrate this point:

Tschirgi (1980) asked subjects to choose the best experiments for identifying which recipe ingredients would result in a "great cake" or a "terrible cake." He found that children predominantly chose experiments that would result in good cakes. Kuhn and Phelps (1982) presented fifth graders with the problem of trying to find out which of several chemical substances were responsible for a reaction that turned a mixture pink. Several children directed their experiments toward trying to produce the pink color instead of identifying the substances contributing to the reaction. Schauble (1990) asked her subjects to figure out which car design features affected the speed of cars. Instead of figuring out which features affected the speed, many children became preoccupied with constructing fast cars.

In another study, Schauble *et al.* (1991) asked their subjects to solve two problems by means of self-directed exploration, one designed to elicit the engineering model — the canal task and the second designed to elicit the science model — the buoyant force task. The canal problem was concerned with the question of how water canals should be designed to optimize boat speed. The children could vary the depth of the canal (shallow or deep), the shape of the boats (circle, square, or diamond cross section), the boat size (large or small), and boat weight (light, or unloaded, versus heavy, or loaded with a small barrel). The canal task was a try-and-see problem with an outcome easily interpretable as being more desirable. The buoyant problem required the children to investigate the effects of buoyant force on objects of different mass and volume. The children carried out experiments by varying variables in the system — object's volume (small, medium, and large); and mass (largest, intermediate, and smallest), and then measuring the extension of the spring with a ruler marked in centimeters. Half of the children began with the engineering problem and then went on to the science problem. The second half of the subjects started with the scientific problem and proceeded to the engineering problem. It was found that the subjects achieved the greatest improvement in strategic performance when they began with the canal task and then went on to the spring task. This, according to the authors, may be due to fact that this order may map more closely into children's natural way of thinking about scientific inquiry.

It is, of course, the aim of science teachers to lead students to possess the scientific model of inquiry. From the above discussion it may appear that the royal way to get children to reach the scientific model of inquiry is to first allow them to engage in activities that encourage them to utilize the engineering model. Such activities may be used as a kind of bridge which might help in decreasing the gap that novices might have between the two kinds of inquiry.

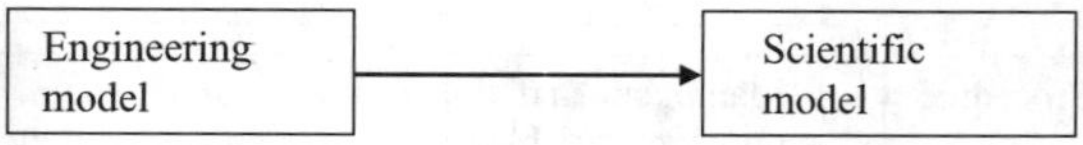

2. Technology-based science teaching is a natural learning environment utilizing cooperative learning. Most educators today will probably agree that cooperative learning should be implemented in the science classroom. This notion is well expressed in reform documents in education. For instance, *The National Science Education Standards* advocate for the use of small student learning groups:

> Using collaborative group structure, teachers encourage interdependency among group members, assisting students to work together in small groups so that all participate in sharing data and in developing group reports. Teachers also give groups opportunities to make presentations of their work and to engage with their classmates in explaining, clarifying, and justifying what they have learned. . . . In the hands of a skilled teacher, such group work leads students to recognize the expertise that different members of the group bring to each endeavor and the greater value of evidence and argument over personality and style. (National Research Council, 1996, p. 36)

Cooperative learning is founded on the belief that student-student discourse promotes cognitive growth and influences students' learning. This belief may be attributed to the *social constructivism* which views knowledge as a primarily cultural product (Vygotsky, 1978, in Windschitl, (2002)). Vygotsky viewed thinking as a characteristic not only of the child but of the "child-in-social-activities" (Moll, 1990, p. 12). Vygotsky's "zone of proximal development" emphasizes the importance of collaborative activities with the notion that the development of a child's mental functions must be fostered and assessed through the assistance of more knowledgeable others.

Based on the literature, Linn and Burbules (1993) suggest the following mechanisms that contribute to effective learning:

(a) Group learning motivates students to persist at a task.
(b) Group learning allows appropriation to occur when students build on someone else's idea to create an idea that they could not have generated alone through, for example, brainstorming.
(c) Group learning can draw on the distributed knowledge of all participants to locate ideas that help construct knowledge.
(d) Group learning provides the opportunity to compare ideas and construct a common point of view. Negotiation of meaning is the crux of the argument for the co-construction of knowledge.
(e) Group learning monitors the progress of students because the tutor or even other members of the group might cue students to check their work, compare solutions, generate self-explanations, or divide a problem into subparts. Furthermore, tutors often reduce memory demands for individuals by keeping track of progress, supplying details that otherwise would need looking up, and prompting helpful behaviors.
(f) Group learning members provide hints or feedback. Vygotsky, according to Linn and Burbules, argued that appropriate hints expand the zone of proximal development and scaffold students as they learn.
(g) Group learning enables the division of the task among group members. The "divide-and-conquer" approach reduces cognitive load for the group and allows the group to accomplish a more complex task.

In spite of the importance and value attributed to cooperative learning, schools attempt to minimize, if not eliminate peer interactions (Duran and Monereo, 2005).

In traditional science lessons, the creation of group tasks requires effort and knowledge. In my opinion, the teaching of science through design and technology activities occurs naturally in groups. When required to design and especially build simple machines, cooperation between students is needed. Students need one another's help when building an artifact. I argue that due to the nature of design and technology tasks, no special effort is necessary.

3. Learning by design utilizes the constructivist approach to learning constructivism. A theory and philosophy or learning that posits, as a result of interaction with the physical and social world, students individually and idiosyncratically construct scientific ideas and beliefs about the world before they receive formal instruction in class. Knowledge, according to constructivism, is always the result of a constructive activity and, therefore, cannot be transferred to a passive receiver.

> If we assume that students have to build up their own knowledge, we have to consider that they are not "blank slated." Even first graders have lived for a few years and found many viable ways of dealing with their experiential environment. The knowledge they have is the only basis on which they can build more. It is therefore crucial for the teacher to get some idea of where they are (what concepts they seem to have and how they relate them). (von Glaserfeld, 1993, pp. 32–33)

Based on a literature review, Windschitl (2002) suggests the following features of constructivist teaching which appear in the left column of the table. On the right column I explain why design and technology activities fit constructive features that appear on the left:

Features of constructivist teaching	**Reasons why the design and technology activities fit the features**
Teachers elicit students' ideas and experiences in relation to key topics and then fashion learning situations that help students elaborate on or restructure their current knowledge.	When raising ideas for designing simple machines students naturally express their own concepts. The evaluation stage of the student's artifact, especially if the artifact does not operate in the manner expected by the student, assists him or her to elaborate on and/or restructure his or her ideas.
Students are given frequent opportunities to engage in complex, meaningful, problem-based activities.	When designing an artifact the student is actually dealing with a real complex problem for which there is no right or perfect solution.
Teachers provide students with a variety of information resources as well as the tools (technological and conceptual) necessary to mediate learning.	The teacher might help the students with their designs by providing them with ideas, presenting similar artifacts to them, or teaching scientific principles relating to the behavior of the artifact.
Students work collaboratively and are given support to engage in task-oriented dialogue with one another.	Design and technology learning environments are natural environments which demand students' cooperation, for both designing and building the artifacts.

continued

Features of a constructivist teaching	Reasons why the design and technology activities fit the features
Teachers make their own thinking processes explicit to learners and encourage students to do the same through dialogue, writing, drawing, or other means of presentation.	First, designing involves the use of drawings. As mentioned before, due to the ability to refer to a concrete artifact, the discourse also includes the use of gestures in addition to words. This, I assume may contribute to the ability of the teacher to express ideas, which are, at least to some degree, explicit.
Students are routinely asked to apply knowledge in diverse and authentic contexts, to explain ideas, interpret texts, predict phenomena, and construct arguments based on evidence, rather than to focus exclusively on the acquisition of predetermined "right answers."	When students design and build artifacts they naturally have to apply their knowledge to the design problem which they are confronted. They, of course, have to predict the behavior of the designed artifact. They also need to evaluate their products based on how their artifact behaved and to suggest alternative solutions to the problems at hand. Of course, there is no right solution in such a problem.
Teachers encourage students' reflective and autonomous thinking in conjunction with the conditions listed above.	Teachers can easily ask students how and why they built their artifacts in the way they did. Did the artifact indeed behave as planned. How did they improve it and why, etc. These kinds of questions may encourage students to reflect on their designed products.
Teachers employ a variety of assessment strategies to understand how students' ideas are evolving and to give feedback on the processes as well as the products of their thoughts.	The artifact itself with the explanation of its behavior, as well as the related scientific rules might provide the teacher with another assessment tool which is currently not accepted by science teachers.

4. *Technology-based teaching promotes question posing*. In his book, *The Disciplined Mind — What All Students Should Understand*, Howard Gardner (1999) writes "On my educational landscape, questions are more important than answers and more important, understanding should evolve from the constant probing of such questions" (p. 24). However, one interesting question is, who's questions should we engage our students with? Brown and Walter (1990) write in their book, *The Art of Problem Posing*,

> Where do problems come from, and what do we do with them once we have them? The impression we get in much of schooling is that they come from textbooks or from teachers, and that the obvious task of the student is to solve them. (p. 1)

Brown and Walter (1990) call for "a shift of control from 'others' to oneself in the posing of problems . . ." (p. 1). They claim that problem posing can help students to see a standard topic in a new light, along with providing them with a deeper understanding of it. They also quote a phrase which, to their opinion, well demonstrates a deep appreciation for the role of problem generating from Chaim Potok's novel *In the beginning*:

I want to tell you something my brother David, may he rest in peace, once said to me. He said it is as important to learn the important questions as it is the important answers. It is especially important to learn the questions to which there may not be good answers. (Chaim Potok, in Brown and Walter, 1990, p. 3)

The importance of question posing dates back to Socrates who wrote "so you will make a law that they must devote themselves especially to the technique of asking and answering questions . . ." (Socrates, in Dillon, 1990, p. 7). This is probably because the ability to pose questions is associated with high order thinking. This is well expressed in the following citation,

Good thinkers are good questioners, taking enjoyment in being doubtful and suspicious of their world, in a positive sense. They take advantage of uncertainty. Why is the world so? Why must it be so? Are other views possible? What other answers might be plausible? Good thinkers utilize questions in particular ways to get at deeper rather than surface meaning. (Hunkins, 1989, p. 15)

Questions, which are essential education tools for all disciplines in general, are of crucial importance in science (Dori and Herscovitz, 1999). As Orr (1999) says, "Good science demands two things: that you ask the right questions and that you get the right answers. Although science education focuses almost exclusively on the second task, a good case can be made that the first is both the harder and the more important" (p. 343). Indeed, the idea of question posing stands at the heart of inquiry-based science teaching. Joseph Schwab (Schwab *et al.*, 1962) who articulated the concept of inquiry-based teaching quite well, envisioned a school curriculum that gave a more accurate representation of the scientific endeavor by practicing scientists, including *active questioning* and investigation. Today, with inquiry-based pedagogy becoming more central with the call of the National Science Education Standards (NRC) that inquiry be a "central strategy for teaching science" (NRC, 1996, p. 31), being aware of children's abilities to ask questions is notably increasing. According to this NRC call, students should learn, among other skills, how to pose a scientific question and to identify and conduct procedures to answer the question. One reason for encouraging and promoting inquiry-based teaching is that children express positive attitudes towards inquiry. Students like to be involved in asking their own questions and formulating ways to answer those questions (Crawford *et al.*, 1999; Gibson and Chase, 2002; Hand *et al.*, 2004).

Despite the importance of children learning to ask their own questions, Dillon, in *The Practice of Questioning* says that children everywhere are schooled to become masters at answering questions and remain novices at asking them. One reason is that teachers are not, unfortunately, properly prepared to teach students how to ask questions. One possible solution from educational researchers, is to offer suitable learning environments to the teachers: environments where children are naturally encouraged to ask questions. I argue that such an environment is the *learning through technology* class. When children design artifacts they naturally start to ask "what if" questions. In addition, when they try out the designs that they build, they naturally start asking "Why doesn't it work?" "How can I improve it?" "Why did the other group's artifact work better?" "What is the scientific explanation for this difference?"

5. *Technology-based teaching promotes systematic thinking*. According to Senge (1990) system thinking is a school of thought that focuses on recognizing the interactions between the parts of a system and then synthesizes them into a unified view of the whole. Furthermore, it deals with recognizing patterns and interrelationships, while learning how to structure those interrelationships into more effective, efficient ways of thinking.

Based on literature review, Ben-Zvi Assraf and Orion (2005), recognize eight characteristics of system thinking:

a) The ability to identify the components of a system and process within this system.
b) The ability to identify relationships among the system's components.
c) The ability to organize the system's components and processes within a framework of relationships.
d) The ability to make generalizations.
e) The ability to identify dynamic relationships within the system.
f) The ability to understand the hidden dimensions of the system.
g) The ability to understand the cyclic natures of systems.
h) The ability to think temporally: retrospection and prediction.

It is important to understand two points: (1) the above attributes of system thinking are not independent of one another, so there may be some degree of redundancy between them, and (2) these characteristics are not necessarily comprehensive.

When trying to recognize system thinking one should not necessarily expect to find all of the above attributes in a given system.

De Vries (2005) points out that the concept of a system can be a strong educational 'tool' to teach about artifacts. According to the author, by making system diagrams of an artifact, its parts (sub-systems) and the way they are connected, pupils and students can gain a first impression of the physical and the functional nature of the artifact. I agree that understanding the concept of a system may help children and older students to understand the artifacts they are dealing with. Learning about artifacts may also help students gain a better insight as to what a system is. This is very important due to the difficulties with which all students of all ages are faced with when dealing with the complexity of a system. For instance, Hmelo, Holton, and Kolodner (2000) found that sixth graders had problems understanding the human respiratory system, partially because they had difficulty understanding the macroscopic as well as the microscopic levels of the entire system. Moreover, they indicated that it is impossible to understand these systems at different levels without understanding the function of the entire system. Kali *et al.* (2003) also reported on students' difficulties in developing system thinking about the rock cycle. It appears that in order to understand how trees function in the forest, it is not enough to understand each tree separately, but rather, to understand how the whole forest functions. Equipping children with systematic thinking, therefore, might help them tremendously with understanding of scientific as well as technological systems. By engaging students with artifacts and encouraging them to deal with questions relating to the operation of the artifact in a system may promote system thinking within children. To achieve this goal, the teacher should expose children to questions such as: What parts make the artifact? Are there any hidden parts? Why are they hidden? How is the function of part 'x' influenced by the function of part 'y'? What will happen if we switch between part 'x' and part 'y'? How will the system behave if only part 'x' is broken?

6. Technology-based teaching encourages the use of thought experiments. In the following, I show that the process of design is associated with thought experiment. Thought experiments, even though entirely the products of mental activity, are viewed as empirical experiments that either cannot or have not been executed empirically,

A thought experiment is an experiment that purports to achieve its aims without the benefit of execution. (Sorensen, 1992, p. 205)

Thought experiments, according to Gilbert and Reiner (2000), "play a major . . . role in science education both by facilitating conceptual change and in relation to some types of practical work" (p. 266). If thought experiments do contribute children's conceptual change, then educators should encourage their students to execute them.

It is my view that thought experiments are crucial in designing tasks. This view is based on the idea that "conceptual construction starts by negotiating meaning, with self and with others, through 'what-if' questions that turn into imaginary experiments in thought, ultimately being applied to the original physical situation" (Reiner and Gilbert, 2004, p. 1821). The following two examples clarify this point:

Example 1: The Parachute Task

In a study examining middle school students learn physics concepts through engagement with simple models, the students were given, among other things, the following design task: "Fill a plastic cap with sand. Now, in groups, design a parachute that will carry the weight so that it reaches the ground in the longest time possible when it is released from a height of 2-meters." The students started to ask questions such as, "What if we had two or even three parachutes instead of one"; "What if we had a big/small parachute"? What if the ropes connecting the plastic cup to the parachute were short/long? From the students' answers it seems as if they ran TEs. The following paragraph is taken from an interview with a student just after he and his team completed building the parachute that they designed:

Interviewer: What did you build?

Student: It is a very novel parachute. It has two covers.

Interviewer: Why do you think this might be a good parachute?

Student: I don't know. I guess it will fall slower. I thought that if with one cover it (the parachute) falls slowly . . . I hypothesize that with two covers it will fall even slower. You see, there are two places that the air can get in [points with his fingers to the upper cover and then to the lower cover and raises his fingers]. The air applies a larger force because it comes in contact with the two covers [the student, again, raises his fingers upwards].

Example 2: The Hot Air Balloon Task

In another task the students were asked to design a hot-air balloon which would achieve the greatest height. A group of students started to work and decided to make a cube shaped air balloon. From what they said to one another it seemed as if they were looking for a symmetric shape. While the group worked, one student sat a bit further off from the group and stared at the sheets of papers which were on the floor. After a while he said:

We need to change the shape of the balloon. It should be extended. [the student meant that the box should be rectangular and not a cubic].

Other student asked him: Why?

The student answered,

At the beginning I thought we needed to make a cubic shape so that we'd get a symmetric shape. But, if it was cubic, the hot air would escape faster from the balloon. If we have an extended box the hot air will have more space to go up and it will lift the balloon up. Also, it will not escape the balloon as fast as in the case of the cubic balloon.

The two paragraphs contain explanations of the designs that the students created. Both explanations are based on concrete details that one can easily use to construct visual representations in his or her head and run mental experiment to test the hypothesis.

In the first description you can easily construct an image of a falling parachute consisting of two covers. You can even "imagine" the air touching the two covers and slowing the parachute down. This, of course, can not be done in reality, since air is invisible. In the second interview concerning the hot air balloon, you can easily imagine an airborne box. The box contains hot air which fills the upper portion of the balloon. Whether or not they are scientifically correct, the children's explanations are very imaginable. This may justify the hypothesis that children may have run experiments in their heads which helped them to test their hypotheses. Based on the results of their TEs they could therein build their parachutes or balloon models. In addition, the students used gestures to clarify their explanations. This too might support the hypothesis that students ran experiments in their heads to test their explanations. Indeed, according to Clement (1994) depictive hand motions are indicators for determining the occurrence of imagistic simulation. From this discussion one may conclude that science teachers may use design activities in their classes, which may encourage their students to run TEs which, in turn, will contribute to the understanding of the relevant scientific concepts.

7. Technology-based science teaching promotes creativity. Although the concept of creativity is an elusive one to define (Hu and Adey, 2002), it is agreed that creativity has a connotation of originality, which may be characterized by novelty, difference, ingeniousness, unexpectedness, or inventiveness (Glover *et al.*, 1989). Sternberg and Lubart (1999), define creativity as "the ability to produce work that is both novel (i.e. original, unexpected) and appropriate (i.e. adaptive concerning task constraints)" (p. 3). According to Boden (1999) novelty may be defined with reference to either the previous ideas of the individuals concerned or to the entire human history. Pope

(2005) argues that this allows for someone to make a discovery or experience a personal break-through (what Boden calls 'P-creative', new to the person), even though it may already be known or have been known at some part in time (in Boden's terms — 'H-creative', new in history). The idea of being creative in reference to oneself has an important role in education since the aim of educators is to encourage and promote creativity within students. Designing, by nature, is described in the following citation, involves innovation of new ideas and transferring them into artifacts.

Engineering design has been defined as, the transformation of ideas and knowledge into a description or artefact, in order to satisfy a set of identified needs; it is the key technical ingredient in producing new products governing the match between products and actual requirements. (Cripps and Smith, 1993, in Court, 1998, p. 143)

In a similar manner, design and technology curricula require school students to generate new ideas, analyze them, make a selection, and describe their artifacts by using verbal and non-verbal representation. Their artifacts should, of course, satisfy a set of requirements. It is thus my understanding, that teaching science through designing may encourage their scientific creativity. Support to the connection of technology and design skills in creativity are items no. 3 and 7 from a *Scientific Creativity Test for Secondary School Students*, developed lately by Hu and Adey (2002):

Item 3

Please think up as many possible improvements as you can to a regular bicycle, making it more interesting, more useful and more beautiful. *For example, make the tyres reflective, so that they can be seen in the dark.*

Item 7

Please design an apple picking machine. Draw a picture, point out the name and function of each part.

According to the authors, this task is designed to measure creative science product design ability.

It is also important to mention that when creative students are taught and their achievements assessed in a way that evaluates their creative abilities, an improvement in their academic performance is noted (Sternberg *et al.*, 1996). Thus, by evaluating their artifacts, students may also gain in achievements and understanding of the science topics. Given the chance to be creative, students who might otherwise lose interest in school instruction, might find that it captures their interest instead (Sternberg, 1999). This is very important, especially in science, which suffers some children's lack of interest. To summarize, it is my view that teaching science through design and technology may be a good idea for improving students' creativity as well as their interest and achievements in science.

8. Technology-based teaching involves bodily knowledge and gestures. I started this chapter by describing the idea of *learning by doing*. I also presented several theories supporting this idea. There is another facet of *learning by doing*. When we *do*, we gain *Bodily knowledge*, which is the kind of knowledge reflected in motor and kinesthetic acts (Reiner and Gilbert, 2000). This knowledge is "stored" in our body and impacts our learning processes. For instance, Clement (1988) showed that embodied intuitions about forces have a role in understanding physics situations. He suggests that knowledge embodied in perceptual motor intuitions is used by experts for physics problem solving. Druyan's work also supports the idea that efficient kinesthetic experiences like jumping training, or measured walking with peers, might help children gain a better understanding of the concept of length. As Duryan puts it "To improve science teaching, teachers are encouraged to be more creative in developing and using active strategies for learning" (p. 1089). In an interesting paper, *Learning With Real Machines or Diagrams: Application of Knowledge to Real-world Problems* (Ferguson and Hegarty, 1995), the authors investigated how learning either from real pulley systems or from simple line diagrams, affected university students' ability to:

a) compare pulley system efficiency;
b) understand mechanical systems; and
c) apply their knowledge to real-world mechanics problems.

In the first experiment there were two learning conditions:

i. Hands-on real condition: The subjects learned by interacting with real pulley systems — they viewed a pair of real pulley systems and acquired information on the system's relative efficiency by actually pulling on the free ends of the ropes.
ii. Diagram condition: subjects learned by viewing diagrams and acquiring information verbally about the efficiency of the systems.

In the second experiment, the authors introduced another condition, the static-real condition. In this condition subjects saw the details of the pulley system configuration but did not observe the motion of the system or experience the weight differences kinesthetically. The experiments showed that subjects who learned hands-on, by manipulating real pulley systems, solved application problems more accurately than those who learned from diagrams. The second experiment showed that it was both the realism of the stimuli and the opportunity to manipulate systems which contributed to this improved performance on the application problems. If the kinesthetic body knowledge contributed to university students' understanding of the physics concepts, for children who most certainly possess lower cognitive abilities at this stage of their life, body knowledge might have an even greater impact on their concept construction. Design and technology activities provide a contact between the child's body and the system. By manipulating the system the child may feel forces, hear, see and smell. This non-verbal knowledge assists the child in gaining a better understanding of the underlying scientific principles fundamental to the system's behavior.

Examples of artifact based science teaching activities

The following examples are of tasks performed both with children and teachers. The results were very similar, but, because the session with the children was not documented, these examples are from the group of teachers.

1. The air car. The first stage consists of presenting the children with an example of a simple air powered car made of two straws connected together and a balloon attached to the end of one of them. Two wheels are attached on the two ends of the straw perpendicular to the one with the balloon, as is shown in Fig. 3.

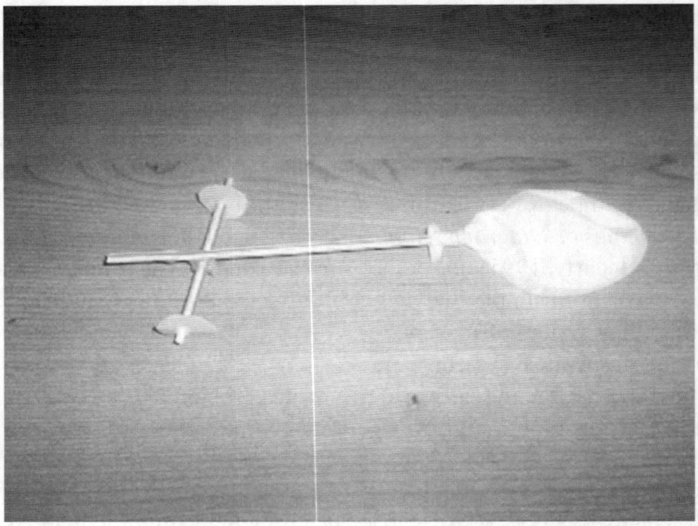

Figure 3. Simple model of air propelled balloon car.

The children are asked what they think the artifact does. The children, seeing wheels on the artifact, will more than likely associate the possibility of movement. They see a balloon and say that it can be inflated. The children can then be asked what they think will happen after the balloon is inflated. Some of the children will then say that it will move much like a car and some may say that it will fly into the air. The children are then asked as to the direction of the movement, supposing it was to move on the ground or in the air.

The children can then take a shot at inflating the balloon and letting it go, observing the movement of the car in the opposite direction of the direction of the air coming out of the balloon.

The concepts which can be learned here are: The balloon is elastic in nature and as a result of it contracting, pushes out the air; the air moves in one direction (the children can try feeling the air flowing out of the balloon) and the car in the complete opposite direction. This is definitely a superb introduction to the teaching of Newton's third law (The balloon pushes the air, which as a result pushes the balloon and with it the car).

The second stage consists of having the children try and improve on the original model. The children can be asked to create a faster car than the one shown to them by the teacher. This, of course encourages work in groups because the children are asked to build something, which is always easier done with the help of another person than by oneself. The children are trying to deal with an open problem where there is no one correct answer. There can be many different approaches to it, all plausible and more than likely to achieve the required goal (the making of a faster model). This opens the door for creativity among the children and allows them to express and use previously acquired knowledge. They can immediately test their ideas as they come to mind, which also encourages question asking: “Why does it move like that”, “Why is it like this?”

The following are examples of refinements made during sessions with children and teachers alike:

a) One group decided to make an axis independent from the main body of the car, meaning that it was able to spin freely. This was done by connecting the wheels to the straw perpendicular to the main straw by use of a toothpick (which could rotate freely inside the straw), thus allowing the car wheels to turn with the propagation of the car. The idea of changing the axis from one that was fixed to the body to one that was more free, led to a discussion on the axis and its function. A discussion also arose on the difference between wheel friction and slipping friction.
b) Another group decided that raising the balloon from the ground (as shown in Fig. 4) by placing it on top of a small water bottle or an aspirin box, would allow for less friction with the floor and therefore also for an increase in velocity. The participants did not limit themselves to the materials shown on the original artifact, but rather chose creative ways of building their artifact using a variety of materials like foamed plastic for wheels or even wheels made of rolls of string, as is shown on Fig. 4.
c) A common factor chosen to increase the velocity of the car, was the number of balloons connected to it. Many of the groups decided to increase the number of balloons from one balloon to two. A discussion was then held on the reasons leading to the increased velocity as a result of adding more balloons, such as increased force and power caused by the balloons. This encouraged a discussion on friction (see Fig. 5).
d) Some of the participants decided that changing the wheels to a smoother material would somehow help increase the velocity of the car. This was particularly interesting as it led to another discussion on the use of the axis — this factor would indeed have a positive effect if there was no unrestrained axis, however much less of an effect when one was present.

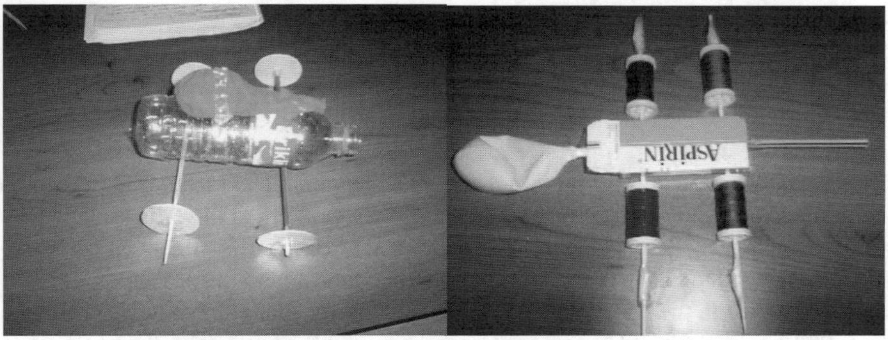

Figure 4. Car design with balloon raised from the ground.

Figure 5. Car design with two balloon for increased propulsion.

e) One especially interesting group decided for some reason to add extra wheels which were not parallel to the original wheels (as shown in Fig. 6). This caused an opposite effect to that which was desired, raising a large number of questions as to why this happened, why it is in fact that when the 2 wheel axis's cross each other they interfere with the cars movement. It also lit up a discussion on how more is not necessarily always better.

2. The parachute. This example was performed on groups of junior high school students. In this example, as was stated earlier in this chapter, the children were shown a simple parachute made of some cloth and strings. Attached to it is a weight of some sort.

The groups were then asked to create a parachute, which takes the longest time to fall, when released from a predetermined height. All the parachutes are given the exact same weights.

The groups try different methods in order to reach the goal, some efficient while others less. Some groups altered the size of the cloth or even tried creating parachutes with two cloths, while others experimented with the impact of different string lengths connecting the cloth.

One group had the misconception that the air slowing the parachute's descent was in the shape of a "pocket." They thought that if they could hold this "air pocket" they would be able to get a considerable increase in the parachute's effectiveness. To do this they decided to take two cloths and place them one on top of another, while making a small hole in the bottom cloth. They hoped that by doing this, the air would go through the first cloth and become entrapped between the two cloths, therefore slowing the parachute considerably. Needless to say, this experiment was a failure and the

Figure 6. Car design with two axis's not parallel to one another.

parachute simply plummeted to the ground. This caused the children to start asking questions as to what caused the parachute to fall so fast and why their experiment failed to succeed.

After all of the groups had finished creating their parachutes, everyone gathered around and discussed which of the parachutes would take the longest time and why. After testing all of the parachutes and gathering the results, another discussion was held as to why some aspects affected the speed of descent more than others. Discussions concerning the force applied to the cloth by the air and how it enables the parachute to slow the descent of the weight. Through this discussion came a discussion on lift force and how it effects the parachute's descent, along with a general discussion on velocity. The effect of different weights, although not tested in the session itself is also discussed and demonstrated.

DISCUSSION

Lee and Songer (2003) argue that even though science has been part of the school curriculum since the turn of the 20th century, there is still controversy as to how school science should be taught in order to deliver the essence of science to students. Moreover, I believe, that educators still struggle with developing teaching methods suitable to children's needs and desires. The aim of this chapter was to discuss the potential that design and technology activities might have in implementing the *learning by doing* approach, hence increasing motivation among the children and their willingness to learn and understand scientific concepts. I started the chapter by describing the notion of learning by doing and explaining how that is supported by the case-based reasoning as well as situated learning theories. I then claimed that though schools make enormous efforts to utilize the *learning by doing* idea, they usually miss the heart of the idea and may in fact detach doing from any meaningful learning. This situation leads to *activity-mania*. There are two reasons for the inefficient implementation of the *learning by doing* approach. The first is the lack of awareness by teachers of the effects that *learning by doing* has on children. The second reason is that teachers themselves lack the knowledge to actually perform *learning by doing*. The present chapter presented the notion that design and technology activities are good vessels for implementing the *learning by doing* approach. This notion relies on the strong association between technology and doing. The following quotation emphasizes this association even further:

> Technology is the practical method which has enabled us to raise ourselves above the animals and to create not only our habitats, our food supply, our comfort and our means of health, travel and communication, but also our arts — painting, sculpture, music and literature. These are the results of human capability for action. They do not come about by mere academic study, wishful thinking or speculation. Technology has always been called upon when practical solutions to problems have been called for. Technology is thus an essential part of human culture because it is concerned with the achievement of a wide range of human purposes. (Black and Harrison, 1992, pp. 51–52)

This association, as well as the idea that the time spent by young children at preschool and early primary school is heavily marked by activity and involves the interaction between the children and physical objects around them, led me to pursue a thorough explanation as to why and whether design and technology may be used to teach science. I came up with the following eight reasons:

1. Children tend to employ engineering models of inquiry rather than scientific models.
2. Technology-based science teaching is a natural learning environment utilizing cooperative learning.
3. Learning by design utilizes the constructivist approach to learning.
4. Technology-based teaching promotes question posing.
5. Technology-based teaching promotes systematic thinking.
6. Technology-based teaching encourages the use of thought experiments.
7. Technology-based science teaching promotes creativity.
8. Technology-based teaching involves bodily knowledge and gestures.

I assume that the above reasons are not the only ones and that the reader may think of other reasons as well. I do hope, however, that these reasons alone will convince the reader that there might be a strong power in teaching science through design and technology. In addition, the chapter provided some examples to help those who may be interested in joining this adventure and progressing it from theory to action.

As was suggested here, I believe that teaching science via technology also helps in overcoming the problem that Edelson (2002) raised regarding the difficulty in making authentic real-world science accessible to children. The author argues that authentic activities that are interesting to students are too open-ended, and require knowledge content and scientific thinking of which students do not necessarily have the base and the means to comprehend. The design and technology activities may, in my opinion, be considered a real-world activity which the child may, with suitable teaching, be able to handle, and that may promote the understanding of scientific concepts.

This chapter dealt with one direction — the use of technology in science. Another interesting question is the one I deal with in the following section.

Should we integrate Science, Design and Technology?

I already mentioned the problem with the design and technology curriculum, in which children that learn technology may become disconnected from the science curriculum. Indeed, according to Barlex and Pitt (2002) there is scant communication between staff in the science and design and technology departments and topics which arise in both curricula may be taught in both subject areas with no connections being made by either teachers or students. This situation, according to the authors, leads to wasted time and the loss of valuable opportunities for enriching children's learning. I also previously argued that by doing so we do not implement the idea that science and technology are part of a seamless web that integrates any distinction. In an effort to try and fix the situation, I herein suggested that the sciences should include design and technology. This approach should, of course, be implemented on top of other methods. One question, which, in light of what has been argued in this chapter, may bother the reader, is whether science, design and technology should be integrated. This is beyond the scope of the current chapter. I only wanted to show that the use of design and technology activities within the science topics has enormous potential in implementing science learning by doing and make science lessons more efficient. I do, however, want to close the chapter by referring to this dilemma. By taking the web-less view into account it might look natural to integrate the two subjects. However, one should seriously consider the argument that Barlex and Pitt (2002) make, according to which integrating science, design and technology is inappropriate. The authors claim that,

> science and design and technology are so significantly different from one another that to subsume them under a 'science and technology' label is both illogical and highly dangerous to the education of pupils. Both are necessary and from their individual positions can enhance each other. Science is essentially explanatory in nature whereas design and technology is aspirational . . . . Design and technology is the area of the curriculum that enables students to intervene creativity to improve the made world. As such it is essential that design and technology is neither deflected from this main purpose nor diluted in its effectiveness by a shotgun marriage. (p. 189)

Does Barlex and Pitt's view contradict the seamless web view? No. It is my understanding that the connection between science and technology can indeed be seen as having a kind of web-structure. However, even in this web one can recognize the technological parts and distinguish them from the scientific parts and vice versa. Although I would avoid using terms such as *illogical* and *highly dangerous*, I agree that integrating the two might cause us to omit some important aspects of each topic. Thus, it is my opinion that each of the subjects should develop its own activities with regards to the other. I suggested that science can develop more design and technology activities which are relevant to science lessons and, on the other hand I also suggested that design and technology might develop scientific activities. In addition, as was also suggested by Barlex and Pitt (2002) I suggested that designers of each topic be aware of the other topic's curriculum so that a better match can be achieved between the two. To summarize my suggestion it might be worthwhile to think of it as islands of technology within the science lessons and as islands of science within the design and technology subjects. The teacher's role would then be to build bridges so that the child can first move securely between the islands and as a result will construct web structured relationships for him or herself.

CHAPTER 4

FROM THE KNOWN TO THE COMPLEX: THE INQUIRY EVENTS METHOD AS A TOOL FOR K-2 SCIENCE TEACHING

In chapter 1, six justifications for science education in early childhood were discussed along with children's capability to think scientifically. So, the first chapter, I hope, convinced the reader that science education should begin at early stages of life. The next two chapters covered a variety of approaches and methods so the convinced reader might encourage teaching science at K-2. All of the previous chapters, however, concentrated on and emphasized the child's needs. The preschool teacher's needs were neither thoroughly nor explicitly considered. This chapter aims to rectify this situation. After all, we should remember that K-2 science education is primarily in the hands of the teacher. A science curriculum, as excellent as it may be and which may fit the children's needs perfectly, might fail because of the teachers. In this chapter I present a fresh idea: science curricula, at the K-2 level, should consider and emphasize the teacher's needs. I do not mean that one should neglect the children's needs: on the contrary, those should always be kept in mind. However, over the years, I have noticed that most curricula are built first from the outlook of the children's perspective, and only then do the designers search for ways to prepare the teacher to implement the program. The common assumption is that curricula should be designed for the children and that at the design stage teacher's needs are not considered. Only after the curriculum is ready is consideration given to its teachers. This, in my opinion, is wrong!

This chapter is divided into three parts. The first part discusses the idea of a curriculum driven by the teacher's needs and presents a teaching strategy that I developed and named *Inquiry Events* (IE), which uses this approach.

The second part describes a research which examined educators' changes in science teaching efficacy beliefs and science teaching outcomes after participating in a workshop on IE. This research also tested the educators' views about IE itself. After realizing that the IE method was well accepted by teachers, I felt it was time to test it in a kindergarten. The last part of the chapter presents a research done with my master's degree student, Liat Bloch, which evaluated the IE in two kindergartens.

PART A: THE NEED FOR A NOVEL TEACHING METHOD — THE INQUIRY EVENTS

Elementary grades have been cited as the weak point of science education (Gardner and Cochran, 1993). In most cases, only a small part of elementary school activities are related to science (Schoeneberger and Rusell, 1986). There is considerable evidence suggesting that K-2 science education worldwide is in a similar state (Mulholland and Wallace, 1996). It is reasonable to assume that even less attention is given to science activities in kindergarten. Two main factors may explain why both elementary and kindergarten teachers have difficulties in being effective science educators: a) teachers possess insufficient scientific background (Franz and Enochs, 1982; Hurd, 1982), and teachers hold anti-science attitudes (Koballa and Crawley, 1985). Shrigley (1974) discovered a low correlation between science content knowledge and teachers' attitudes toward science. These results suggest that addressing only the problem of insufficient knowledge and requiring additional science oriented courses, namely through mathematics and science departments, as part of pre-service and in-service teacher preparation, may not be the most appropriate solution to training competent elementary school science teachers. Moreover, I concur with Tosun (2000) that such courses may have a counterproductive impact on teachers.

Wallace and Louden (1992), wondered: "Why, after more than three decades on the reform agenda, elementary science teaching continues to disappoint. Is it because we haven't found the right 'formula', or could it be that we have an imperfect understanding of the problem and unrealistic expectations for the solution?" (p. 508).

Teachers gravitate toward tasks where they feel confident and competent (Cunningham and Blankenship, 1979). Although children's abilities and interests must be one working assumption while designing a curriculum, another is that science curricula must also consider teachers's interests and abilities. One such approach is called the 'Inquiry Events' (IE) teaching method. This method involves dealing with open-ended problems taken from real-life situations, encouraging investigating different kinds of issues (ethical, economic, aesthetic, etc.) which teachers at both kindergarten and elementary school consider and discuss. The method helps teachers to introduce scientific questions relating to those daily situations, which they would normally ignore or omit. The assumption is that by combining science with familiar day-to-day situations, teachers will feel that scientific knowledge has practical importance in everyday life. Moreover, since the scientific questions are only part of the whole problem, once the teachers feel confident, it is believed they will be more willing to go further to gain additional and necessary scientific knowledge. In this manner, it will be easier and more natural for a teacher to include scientific questions among other issues arising from a concrete real-life problem, rather than focusing separately on a problem which is somewhat fictitiously defined as a *scientific* one.

Stages of IE Design

Developing an inquiry event includes two main stages: choosing an appropriate inquiry event that fits some criteria which will next be described, and then expanding the inquiry event into suitable learning units. Fig. 1 illustrates these two stages. The upper part — part A, shows the first stage of the IE development. The lower part — part B, shows the second stage.

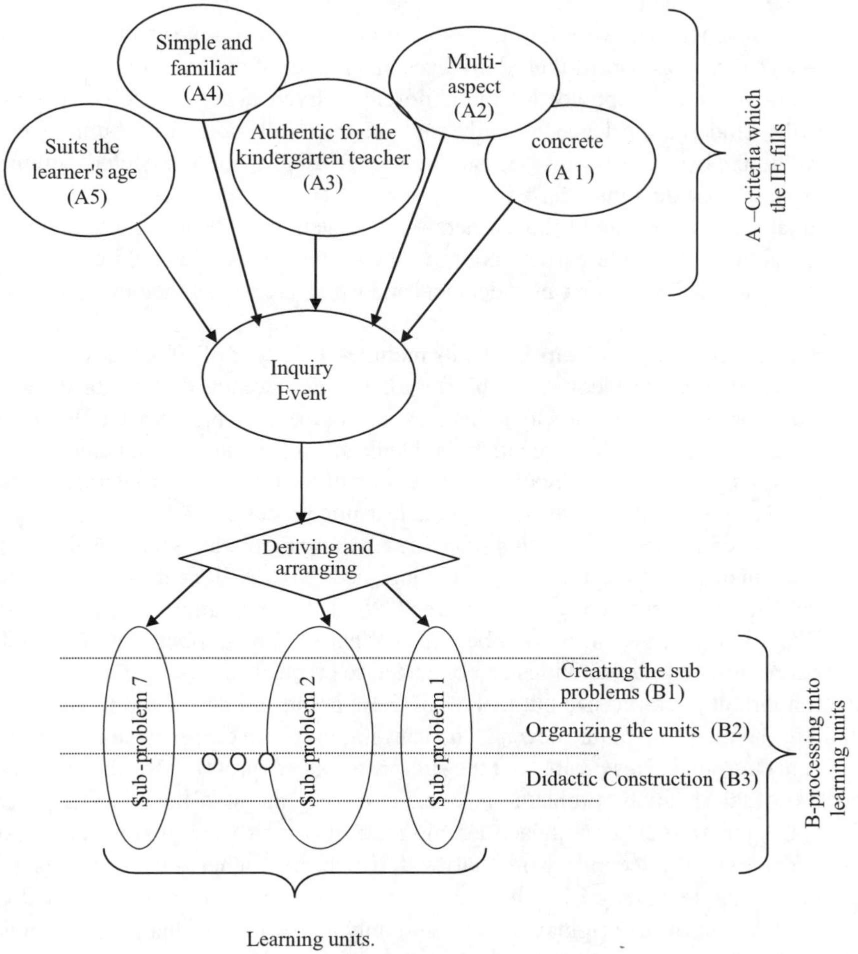

Figure 1. The two stages of IE development: The first stage — stage A, choosing inquiry events according to the five detailed conditions. The second stage, stage B, processing inquiry events to learning units.

Stage A: Criteria which the IE Fills. For an open ended problem to be considered an Inquiry Event, it should fulfill the following criteria:

1. Concrete: The IE should deal with a real and concrete, open-ended problem taken from a real-life situation. In addition, the event is dealt with in a veritable manner, and the learning group is asked to gradually reach a solution.
2. Multi-aspects — The problems encourage investigating different kinds of issues—ethical, economic, aesthetic, geographic, scientific, technologic, etc., — meaning that scientific questions should only be *part* of the whole problem.

3. Authenticity for the kindergarten teacher — The intention here is that kindergarten teachers deal with familiar events in their daily routines and problems that they deal with as part of their daily lives, in and out of the kindergarten.
4. Familiarity and simplicity for the children — Even if the events are to be part of the kindergarten teacher's daily routine, they still have to be familiar to the children. They also have to be simple and feasible to be conducted within the constraints of the kindergarten.
5. Suitability for the age of the learners — The event must have content and activities suitable for kindergarten children. The scientific issues should especially be ones which K-2 children can understand and which are part of their every-day lives.

Stage B: Processing into Learning Units includes a number of sub-stages:
1. Creating the sub-problems — In this stage the inquiry event is disassembled into secondary problems connected to different aspects of the main problem, the IE. The purpose is to turn each of the secondary problems into a separate learning unit.
2. Organizing the units — Creating a succession of the derivative problems, to enable a logical and suitable progression of the learning process.
3. Didactic construction — In this stage the teaching methodology is developed, and questions such as the following are addressed: How will each of the secondary problems be presented to the children? What demonstrations should be used? What kind of experiments will be done? What kind of artifacts can the children build? How will the activities be connected to create a succession?

It is important to emphasize out that all of these points are developed relating to the main problem — the inquiry event. The learning units are designed so that the IE is in the background. This means that the sub-problems are presented to the child in the context of the IE's main problem presented at the beginning of the teaching process. The children are reminded in each learning unit that their final goal is not just simply to deal with the secondary problem with which the teacher is currently teaching, but rather that the main goal is the IE. For this reason, most of the sessions will start with a reminder of the original problem and only afterward will the sub-problems be presented along with their connections to the main problem.

How and Why IE differ from Other Teaching Methods

I would also like to point out and emphasize the differences and similarities with IE and similar pedagogical methods such as problem-based learning and project-based learning. Both problem-based learning and project-based learning were discussed in the second chapter. As previously explained, the underlying idea of problem-based learning is that learning and teaching processes are driven from and start from a problem which is presented to the learners and they are required to deal with it. However, problem-based learning is a very general term which might include a variety of teaching activities. The problem the teacher selects may be narrow or broad; it can be specific or general; and it may be one which requires a small or a large portion of time to complete. Further, no constraints are imposed on what rules the problem should obey. The IE method begins the teaching/learning processes with a problem and hence it belongs to the PBL approach. However, not every problem can be considered as an IE, but rather only these which obey the constraints that were presented earlier. It should not be too short or too long. The IE should last about 7–8 weeks and cover different aspects of the IE's main problem. One may say that the IE is a kind of a project and thus the same old lady under a different cloak. I agree that the IE method in some sense is similar to a project. However, there is a big difference. First, in projects, the learners themselves usually bring the problem that they want to pursue. It is not a systematic pre-designed curriculum. On the contrary, the child may bring in a project on a topic which the teacher knows nothing about. In the second chapter I criticized such an approach. It is my opinion that most pre-school teachers do not have the knowledge, skills or means to present or use such a method efficiently. The IE, as opposed to project-based-teaching, is a systematic curriculum based on problems with which the teacher is familiar. In addition, projects may include many different aspects. But there is no such demand that these aspects all be included, and may deal with only one aspect that the child chooses to concentrate on. Moreover, even if it does involve several aspects, it might include non-scientific topics. For instance, a child can choose to start a project on water. His or her project might include biology aspects, physical aspects, and chemical aspects, but there are no requirements for it to include non-scientific topics such as literature or history. The IE approach requires that a problem include several aspects, including non-scientific ones. For these reasons, I find the IE method to be different from other problem-based teaching methods.

An Example of IE: The Friend Abroad

A friend of the students in a K-2 class has left Israel with his entire family and now lives in England. He would probably enjoy receiving videotapes and books in Hebrew. The problem: sending a parcel which would include videotapes and books as well as some other goods. Several questions arise from this situation, relating to the problem at hand: Ethical questions (should we send him a package?); Geographical questions (where is England? How far it is from Israel? What language is spoken in England?); Economic questions (how much money would it cost to send a package containing several videotapes, books and some candy? How much money would each of the students need to contribute?); Technological questions (what materials should we use for the package? How would we send the package?) Scientific questions (What is weight? How can we measure the weight of the package? What kind of candy is appropriate to put in the package in regards to melting? What is melting and how does it happen?).

In the above IE example it is reasonable to assume that a K-2 teacher would be familiar with the situation. Teachers have certainly sent parcels at some point in their lives. They are familiar with the connection between the cost of sending a parcel and its weight. They know how to weigh. They are also aware of the things that should be put in a parcel and things that should not. So, with very little help, the teacher may understand the relevant underlying physics principles required for dealing with this IE. The ethical aspect is probably one in which the teacher is an expert and knows what

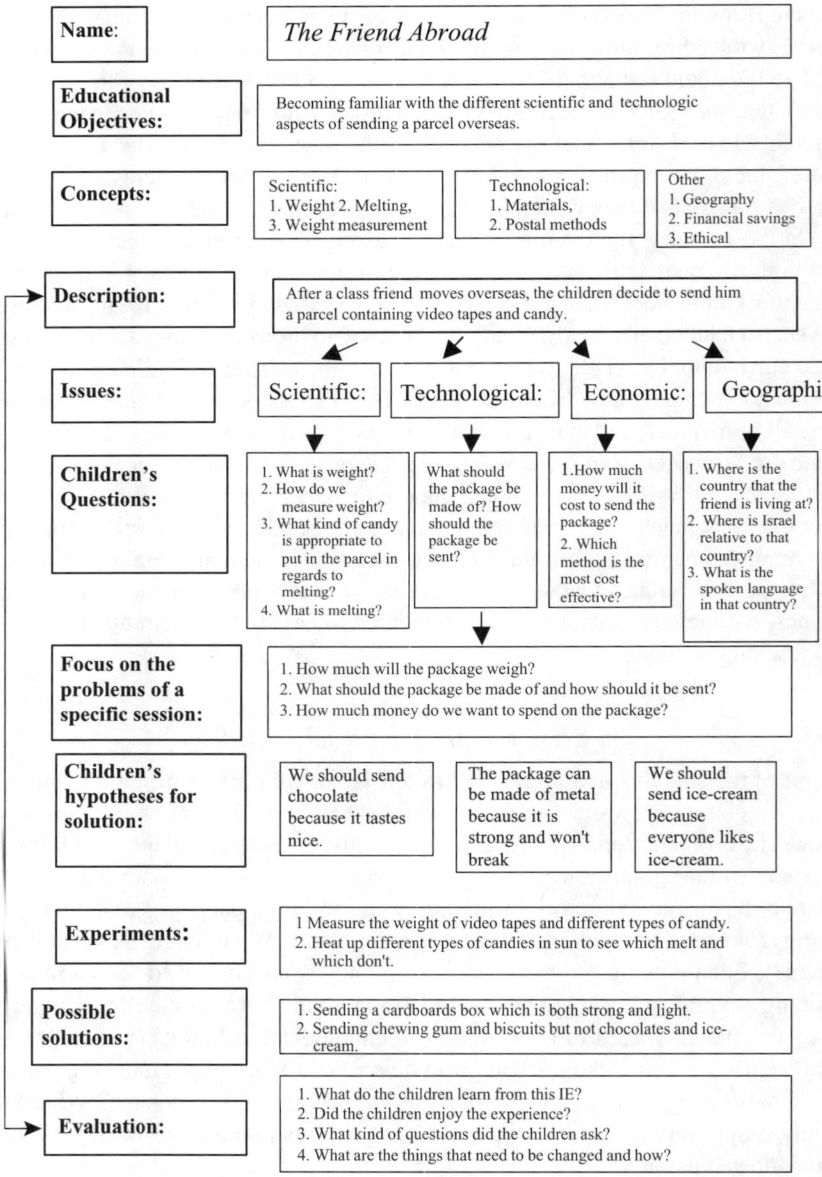

Figure 2. The use of the IE design instrument (IEDI) for the Friend Abroad IE.

and how to teach to his or her children. The mathematical aspect is also relatively simple for the teacher.

As can be seen from the example, unlike the project-based approach, which might demand that teachers deal with problems with which they have minimal knowledge, and therefore might require them to devote much time to the learning of that subject, the IEs must be familiar to the teachers. Whenever I present the idea of IE, there are always educators who approach me after the class and tell me that they were surprised at how much science could be taught through such simple daily situations which they could implement in their classes. Indeed, the IE method can be seen analogically as a flashlight which sheds its light on the scientific issues of problems which teachers have already experienced. To help teachers design their own IE the Design Inquiry Event Instrument (DIEI) was developed as shown in Figure 2. DIEI guides the teacher through the necessary stages of designing IE: naming the IE, detailing its objectives, indicating the main concepts, describing the IE story, raising different IE questions, focusing on difficulties, hypothesizing children's possible answers, suggesting relevant scientific experiments, and offering possible solutions for the problems. Figure 2 is a DIEI for the example that was provided in this section.

PART B[1]: INQUIRY-EVENTS AS A TOOL FOR CHANGING SCIENCE TEACHING EFFICACY BELIEF OF KINDERGARTEN AND ELEMENTARY SCHOOL TEACHERS

Elementary School Teachers' Beliefs and Attitudes toward Science Teaching

Although there are few previous studies specifically addressing the question of anti-science attitudes among elementary school teachers, literature suggests that such attitudes do exist (Gustafson and Rowell, 1995; McDuffie, 2001; Parker and Spink, 1997; Skamp and Mueller, 2001; Stepans and McCormack, 1985; Tosun, 2000; Yates and Chandler, 2001). Like most people (Nemecek and Yam, 1997; Park, 2000), it may also be that many teachers regard science merely as a school subject detached from everyday life. Indeed, elementary school teachers are not known to be science oriented (Cobern and Loving, 2002).

Motives and interests are influenced by our attitudes (Miller *et al.*, 1961). Teachers who feel this detachment from science would, at best, regard teaching it as simply fulfilling an obligation (Cobern and Loving, 2002). Given the tremendous impact they have on children, and on the success or failure of any curriculum, teachers' knowledge of science and their attitudes toward it, should be of significant concern. A more fundamental factor, often discarded in existing literature, which may explain elementary school teachers' behavior toward science teaching, is their belief system.

"Belief" is "information that a person accepts to be true" (Koballa and Crawley, 1985, p. 223), differing from "attitude" which is "a feeling, either for something or against it" (Miller *et al.*, 1961). Attitudes stem from beliefs, and both are related to behavior (Riggs and Enochs, 1990). The following example, taken from Koballa and Crawley (1985) will clarify this point. An elementary school teacher judges his/her

[1] This part is based on a paper that was published in *The Journal of Science Education and Technology* 12(4): 495–502.

teaching ability to be lacking in science (belief), consequently developing a dislike for science teaching (attitude). The result is a teacher who avoids teaching science if possible (behavior). Teachers' beliefs can be divided into outcome expectancy beliefs, and self-efficacy beliefs. "Teachers who believe learning can be influenced by effective teaching (outcome expectancy beliefs) and who also have confidence in their own teaching abilities (self efficacy beliefs) are more persistent, provide a greater academic focus in the classroom, and exhibit different types of feedback, than teachers who have lower expectations concerning their ability to influence student learning" (Gibson and Dembo, 1984, p. 570).

OBJECTIVES

The goals of this research were:

1. To examine changes in the educators' science teaching efficacy beliefs and science teaching outcome expectancies, as a result of a four-day workshop based on Inquiry-Events;
2. To examine teachers' attitudes, resulting from a four-day workshop, toward the Inquiry Event method as a tool of teaching science in kindergarten and elementary school.

METHOD

In the spring of 1997 and in the spring of 2001, two four-day-workshops on science for kindergarten and elementary school teachers were conducted[2]. The first group (spring, 1997), consisted of 30 participants, while the second group (spring, 2001), consisted of 28 participants. Among the participants were experienced K-2 teachers, curriculum developers, and teaching-trainers from 20 different developing countries in Asia, Africa, Eastern Europe and the Caribbean Islands. The IE related workshops each lasted four days, for about 9 hours each day. The workshops included an opening lecture explaining the meaning of IE's, observing IE activities in a kindergarten, and participants designing IE's in small groups (3–5 participants) using the Design Inquiry Event Instrument (DIEI), whose framework is shown in Fig. 2. The IE's designed in each group were then presented to the entire class, and were followed by a discussion.

The scientific questions of the IE's that were presented to workshop participants related to scientific concepts such as weight, temperature and light, and included demonstrations and active experimentation.

The Science Teaching Efficacy Beliefs Instrument (STEBI) was chosen to be used in the pre-test and post-test, since it was found to be a valid and reliable tool for investigating elementary school teachers' beliefs toward science teaching and learning

---

[2] The workshop was the first part of a 20 day course on "Science Education in Early Childhood" organized by the Golda Meir Mount Carmel International Center, in the framework of MASHAV, The Center for International Cooperation, Ministry of Foreign Affairs, Israel.

(Riggs and Enochs, 1990). The statements in the STEBI questionnaire are assembled into two distinct item categories:

1. Personal belief toward science teaching efficacy- for instance: "I am continually finding better ways to teach science"; "I do not know what to do to encourage students to take an interest in science."
2. Outcome expectancy belief of science teaching- for instance: "When a student does better in science than usual, it is often because the teacher exerted some extra effort"; "Even teachers with good science teaching abilities cannot help some kids learn science."

To analyze the influence of the IE method on designing science activities, questions were added to the STEBI questionnaire. In the pretest, the participants were asked to describe the most stimulating science activity they had conducted in their class prior to the workshop, and in the post-test they were asked to design a similar activity, that may contribute to the development of children's cognitive skills. They were also asked to give their opinion in the post-test, regarding the IE method and its potential as a tool for science teaching.

RESULTS

Beliefs toward Science Teaching

Pre-test and post-test means and standard deviations for each of the two dimensions of science teaching efficacy belief, are shown in Table 1, which indicates that the means increased in both categories. The results of the t-paired sample test shown in Table 2, indicated that the changes in both dimensions of the participants' teaching efficacy beliefs were statistically significant.

TABLE 1. Means and Standard Deviations (SD), for the two Dimensions of Science Teaching Efficacy Belief

	Pre-test		**Post-test**	
	Mean	**SD**	**Mean**	**SD**
Personal science teaching efficacy belief	3.45	0.38	3.94	0.37
Science teaching outcome expectancy	3.95	0.13	4.45	0.19

TABLE 2. Results of t-paired Sample Tests for the two Dimensions of Science Teaching Efficacy Belief

	t	***df***	***p***
Personal science teaching efficacy belief	2.23	57	<0.05*
Science teaching outcome expectancy	2.15	57	<0.05*

*Significant at the 0.05 level.

Teachers' Beliefs Regarding the IE Method

In order to examine teachers' beliefs regarding the IE method as a tool for science teaching in K-2, the participants were asked to express their opinions in the post-test questionnaire. All of the participants felt that the IE method is an efficient tool for teaching science, especially in early childhood. The reasons they gave were divided into the following categories: (1) IE presents science to a child as an integral part of life; (2) IE helps to teach science effectively; (3) IE contributes to the development of the child's cognitive skills; and (4) IE helps children develop social skills. Table 3 demonstrates some of the participants' statements, according to the four categories.

Changes in Perspectives regarding the Most Stimulating Scientific Activity

Changes in teachers' responses in describing the most effective science activities in kindergarten and elementary school in pretest and post-test were compared. It was

TABLE 3. Categories of Participants' Beliefs about the IE Teaching Method

Benefits of IE Method Mentioned by Participants	**Samples of Teacher's Answers**
IE presents science as an integral part of life to a child	"It helps children to see many realms of life as a whole." "It helps integrate science with other subjects, and does not leave science as an isolated issue in children's minds." "It relates science to real life situations."
IE helps teach science effectively	"It develops teacher's creativity, by releasing them from conventional teaching styles . . ." "It facilitates hands-on experiments." "It enables the teachers to plan their objectives (be it general or specific)."
IE contributes to the child's cognitive development	"It helps children to organize their thinking processes and arrive at logical solutions." "It helps accepting and remembering ideas." "It helps a child to solve practical problems." "It helps children to develop cognitive skills, because they find solutions from different alternatives on their own." "It stimulates the children's thinking, rather than only giving information."
IE helps children develop social skills	"It helps develop communication skills." "It gives the children an opportunity to work in groups and develop team work skills."

found that in the pretest, 10 of the 58 participants (17%) gave no answer at all. Thirty-one participants (53%) mentioned science activities entailing only observation and description skills. Thirteen participants (22%) depicted activities involving the use of categorization skills, and only four participants (7%) described activities requiring problem solving skills. In the post-test, eight participants (14%) described activities involving only observation and description skills, 28 participants described IE's (48%) and 22 participants (37%) depicted activities requiring problem solving skills which are not IE. The results indicate a notable increase in the design of science activities requiring problem-solving skills.

IE's Designed by Participants

One of the tasks given to the workshop participants was to design their own IE's. Some examples were: 1) Going to the Beach, 2) Owning an Aquarium, 3) Raising a pet animal, and 4) Visitors in the Kindergarten. The IE helped the participants to come up with scientific and technological questions requiring the use of problem solving skills. The participants indicated that, although raising a pet animal in kindergarten is a familiar experience to most of them, they were not aware of its potential to develop the child's problem solving skills. Prior to the workshop, they regarded this activity only in the context of developing the child's observation and verbal skills.

SUMMARY

It was found that IE is a highly efficient teaching method in our search for a more effective approach to improve science teaching efficacy beliefs of kindergarten and elementary school teachers. This investigation also indicates that significant changes in teachers' belief systems toward science teaching can be produced in a short period of time. These results corroborate with those of Spooner and Simpson (1979), who found that a significant change in teachers' attitudes may be achieved within a few daily sessions.

This study also indicates that during the workshop, teachers acquired positive attitudes regarding IEs. An analysis of teachers' beliefs toward IE's signified the potential of this tool to promote teaching science in kindergarten and elementary school and to contribute to the development of children's cognitive skills. The participants predominantly mentioned that the greatest potential of IE's is its ability to introduce science to a child as an integral part of life, and not as an isolated problem. This is consistent with Dewey's approach (Dewey, 1916) to the teaching process, which requires taking the psychological needs of a child into consideration, rather than introducing science as a logical coherent subject. Dewey's approach was further discussed in chapter B. According to this approach, teaching processes should be based on the child's mundane experiences, practical day-to-day problems and familiar intelligible issues, which are of vital interest to the child. In my opinion, IE may be an appropriate tool to organize the teaching processes in the necessary "psychological order."

In addition, it was found that the teachers' point of view regarding science activities had been changed as a result of the IE method. Prior to the exercise, the most stimulating science activities described by participants required mainly the use of observation and verbal skills. At the end of the workshop, most participants agreed that the most stimulating science activities involved problem solving skills. Interestingly, some participants denoted that using the IE method influenced their own thinking processes: in the words of one of the participants: "it has made me a better thinker." However, most participants noted that the IE "calls for a lot of hard work," and its design requires more practice as well as additional training courses.

This study is preliminary in its nature. Other studies need to be designed, to replicate this treatment with a larger number of subjects from different educational backgrounds and realms, as well as to attempt to answer other questions that did not arise in this research. The next part of this chapter discusses implementing IE's in kindergartens.

PART C[3]: BRINGING INQUIRY EVENTS TO THE KINDERGARTEN: INQUIRING INQUIRY EVENTS IN THE FIELD

In part A of this chapter the general idea of IE and the rationale for using it were presented. Part B described the impact of IE on educators' science teaching efficacy beliefs as a result of a four-day workshop based on Inquiry-Events. This part continues examining the IE approach. The study presented here evaluated the IE teaching method in two kindergartens which made use of the *inquiry events* method. Specifically, the following questions were addressed:

- What was the kindergarten teacher's point of view toward science teaching in kindergartens before learning about IE?
- What was the kindergarten teacher's point of view toward implementing the *inquiry event* method?
- Is, and how is the program suitable for the kindergarten teacher's needs?
- What teaching methods were employed by the teachers?
- Were the IEs suitable for children?

METHOD

Participants

The study included two kindergarten teachers and the children in their kindergartens. Both kindergartens were around the city of Haifa: a kibbutz kindergarten where, at the time of the study, 30 children attended from ages 4 and 8 months to 6 and 5 months, and an urban kindergarten with 26 children from the ages of 5 and 5 months to 6

[3] This part is based on work which Liat Bloch carried out as part of her master thesis and I thank her for allowing me to use the name of her thesis as the name of the chapter.

and 8 months. The kindergarten teachers were certified kindergarten teachers. The teacher from the urban kindergarten had 10 years experience, whereas the kibbutz teacher had 12 years experience.

Description of the Kindergartens

The urban kindergarten is situated in the heart of a noticeably poor urban neighborhood. The houses are four stories high, there is no greenery surrounding them and there are no gardens. The population appears to consist primarily of lower class immigrants coming from all types of different places. The kindergarten itself is an old structure which, at the time of the study was scheduled to be demolished and rebuilt in a more successful manner. The structure of the shops in the vicinity of the kindergarten displays the general impression that the area serves the surrounding community. The shops seemed active but very shabby. The picture inside the kindergarten and in its playground contrasts the surroundings that were just described. The kindergarten is small and packed but is full of materials and has an active feeling in it. The walls are decorated and the children look busy to the random visitor. There is a feeling of everlasting nurturing and investment in everything. The kindergarten playground enjoys a natural grove, full of scraps and different fixtures. It also spreads out on a vast amount of land. The pre-school teacher described the playground as one that suffers from vandalism from bored youths of the neighborhood. The kindergarten group of children is described by their teacher as one that comes from a weak background and she feels that the kindergarten fills many voids within them.

The kibbutz kindergarten is fundamentally different in nature. The feeling that welcomes the visitor is one of space. The children's freedom of movement is evident in the huge grass playground. They stop to pick strawberries from the strawberry bush on their way. The structure of the kindergarten is that of a conventional kibbutz kindergarten: The yard is rich with scraps. The area used for activities inside the kindergarten is very large and is divided into different corners. The meeting is not part of the playground, so the children can continue what they were doing while the staff organizes it. This contrasts with the urban kindergarten where there was little space left while class preparations were taking place. Another noticeable difference is the meeting with a broad staff. The kindergarten teacher and nanny are constant figures. However, in all of my visits, there were two additional grownups helping with the work. The kindergarten consisted of children from the kibbutz and in addition, children from the surrounding area like Haifa and surrounding settlements, where parents were interested in having their children brought up in this type of framework. The group of children is considered a strong one by the eyes of the kindergarten teacher: it consists of children that come from a nurturing environment.

This study is not meant as a comparison between the two completely different children populations. Comparing the IE method in two kindergartens with such distinct socioeconomic differences between the groups of children, will help assess the applicability and suitability of this and method in a wider variety of kindergartens.

Procedures

The study included the following stages: (1) Locating willing kindergarten teachers suitable for participation in the study; (2) Obtaining entrance permits to the kindergartens; (3) Preliminary work with the kindergarten teachers: specifically the program, finding a suitable IE for them, and constructing a work schedule where one or two weekly sessions could be integrated with their current schedule; (4) Conducting the program in the kindergartens: both having the teachers learn the program and conduct the sessions; (5) Gathering the data: teaching the teachers, videotaping the lessons, documenting the conversations with the kindergarten teachers during visits to the kindergartens, and conducting interviews with the teachers; (6) Rewriting or revising material; and (7) Analyzing the data.

The first five stages are field work and were conducted between April and August of 2003. The observations took place during May and June of the same year, on the days in which the sessions took place according to the program. The interviews took place at different dates during the study.

Tools of the Study

The IEs

1. The IE in the kibbutz kindergarten — "A Guest Visits the Kindergarten." The future event that will occur is the visit of a guest to the kindergarten. The problem the children are confronted with is to prepare for the visit. The problem is real in a sense that the children make real preparations and at the end of the process the guest will arrive. The other conditions are also fulfilled as this is a multi-aspect problem, familiar, simple and well known to the kindergarten teacher. Secondary problems extracted from the main problem include: drink preparation and creating napkin dispensers. These problems are structured to teach scientific and non-scientific topics, for example: Drink preparation is used to teach the topic of liquid concentration. The resulting learning units are organized so that the first is dedicated to presentation of the problem and planning the preparation. The following units deal with the preparations themselves. These units are conducted through group sessions which are then followed by personal experience. The program details realizations that will be implemented in discussions and different activities during the hands-on experience portion, including napkin dispenser construction (3rd unit), which constitutes an opportunity to build a technological artifact. The last unit is the actual hosting and constitutes the conclusive unit of the entire IE.

2. The IE in the urban kindergarten — "Sending a Parcel." This IE is similar to the "friend abroad" IE. Much like in the "Friend Abroad" IE, the Sending a Parcel IE presents a real problem: The children pack the parcel and send it (unit 6) to the children of the "Nitzan" kindergarten, with whom they had contact during the school year. This is an example as to how the original IE, the Friend Abroad, can be changed to fit the real experiences that the teacher and the children had. The IE's problem is presented in the opening unit: We wish to send a package to the kindergarten we visited. Here, the problem is also simple, familiar to the children and the kindergarten teacher, and solving it is real. The next units deal with choosing the appropriate contents. One of the scientific topics which was pursued in some of the units is the topic of weight. The following routine took place in most of the lessons: First, there was an assembly of all the children. During this stage the kindergarten teacher reviews what has been learned and starts a discussion on the new subjects through use of realizations provided by the program. For example, during a hands-on session of the program, the children experienced using the balance scale they were learning about. During the second lesson they filled plastics cups with a sand-like material or gravel and compared the two using the scale. On other days, different aspects of the program were emphasized, which gave the children a chance to experience them first hand, for example: decorating. Building a technological artifact was presented here as well (during unit 6), where the children built personal scales. The activity stage in the kibbutz kindergarten usually involved children working simultaneously around a number of tables; in the urban kindergarten, activities were conducted the around one table while the children took turns according to the teacher's instructions.

Observations

Observations by one of the researchers (L. B.) took place in both kindergartens for the entire working days that the IEs were executed. Thus, the researcher could obtain more information, especially about the atmosphere in the kindergarten before and after the IE's activities. There were 15 such days. Eight observations were conducted in the kibbutz kindergarten and seven in the urban — one observation for each session. The lessons were also videotaped.

Concluding Session

The last session in each of the kindergartens is a concluding session which is not necessarily required by the program. This session functions as a sort of posttest, which aims at testing the knowledge gain of the learners at the end of the study period. Questions about the studied topics were prepared for this session.

1. *Questions for the Sending a Parcel IE:*
    a) Should I, in your opinion, include tomatoes when sending a parcel? Should I include biscuits? How can I know that the object I chose is suitable for sending in a parcel?
    b) Why is it preferable to pack the parcel in a cardboard box and not a metal one?
    c) Why are there differences between the weight of a stone and cotton wool?
    d) Why is gravel heavier than foamed plastic?
    e) How do we use the scale to measure how much a parcel weighs?
    f) Why is it important to know the weight of the parcel?
2. *Questions for the Guest Visits the Kindergarten IE:*
    a) How many jugs of raspberry juice to be put on the table did you prepare? Why in fact did you use four jugs?

b) What affects the sweetness of the raspberry juice? What makes it sweeter and what makes it less sweet? Why?
c) If I have a glass of raspberry juice which is too sweet for me. What should I do to decrease its sweetness?
d) What is the napkin dispenser made of? Is there a reason for choosing wood for the dispenser? What affects the stability of the wooden dispenser?
e) How would you recommend measuring height when making a personal card?
f) If I wanted to buy a tablecloth, how would I explain to the salesman which tablecloth I want? What is the difference between measuring the height of a child and measuring the table? Why is it different?
g) What happens to a balloon when it is inflated? What happens to the balloon when it is inflated and grows until bursting? Why is a soccer ball harder when it is inflated than when not inflated?
h) What happens to the chocolate when making fondue? Why?

Interviews

Semi-structured interviews were conducted with the kindergarten teachers on the following subjects: (1) The general view of how the kindergarten teacher sees science teaching in kindergartens. (2) The teacher's view concerning the IE method. (3) General impression of the learners during the program.

Field List

Since L. B. stayed in the kindergarten beyond the time during which the program was conveyed, she could have random conversations with the teachers, with the nannies or assistants, as well as the supervisor (on the day she visited) and with the children before and after the sessions. These conversations were documented and used later as additional data.

ANALYSIS

The data can be divided into two main parts: (1) Teachers' views concerning science teaching in kindergartens and the IE method; and (2) teaching processes that took place during the IE lessons. The views were mainly summarized from the interviews and the random conversations conducted with the teachers on the different occasions. The teaching strategies were identified through inductive analysis (Patton, 1990) performed on the observations, which were transcribed verbatim. This includes extracting from the data patterns, themes, and categories of analysis. It is done by the following procedure: The transcriptions are first reread by each of the researchers individually to formulate a tentative understanding (Roth, 1995). The data was then organized, again by each researcher separately, to search for patterns that describe and demonstrate teaching processes and strategies. In subsequent readings, we attempted to confirm the tentative understanding of the phenomena on the tapes. Initial categories of teaching processes were established. Then, as part of the verification methodology (Strauss, 1987), the two researchers repeatedly re-read the data together. Initial categories created separately were revised as a result of several rounds of discussion. Three final categories were finally established:

1. Strategies advancing scientific knowledge — this category was sub-divided into four sub-categories: (1) Familiarization of new terms by announcing part of the term — this sub-category included statements where the teacher announced part of the new scientific concept and then waited for the children to complete it. (2) Explaining the new term by referring to its verbal meaning — this sub-category included all the statements where the teachers used different forms of the word to clarify its meaning; for example: Concentration, concentrated, concentrate. (3) reinforcing understanding by purposely referring to a wrong possibility — this sub-category included all statements where the teacher purposely presented incorrect uses of scientific terms or explanations, and (4) using analogies to reassure understanding — this sub-category included all of the statements where the teacher made use of analogies to clarify scientific phenomena.
2. Strategies advancing scientific reasoning — This category was divided into the following sub-categories: (1) directing to specific features in observations — this sub-category included statements where the teacher directed the children to search or take note of specific features or properties within the object that they were dealing with; for example: its size, shape or structure; (2) advancing causal thinking and thinking in a multi-constrained environment — this sub-category included statements where the teacher or the pupils drew connections between variables, or took different constraints into consideration; (3) encouraging the drawing of generalizations — here we included the statements where the teacher encouraged the drawing of generalizations, or where the children made such generalizations themselves.
3. Strategies used to recruit children's attention and advance coherent understanding of the IE — This category was divided into the following two sub-categories: (1) brief reminder — this sub-category included all the statements where the teacher reminded the children of the main problem in order to pursue the IE activities; (2) encouraging meta-cognition — this sub-category included all the statements where the teacher encouraged the students to think on their thinking by asking question along the line of: How did we come to that conclusion? What did we do? How did we come to know?

RESULTS

Views of the Teachers

Teachers' Views concerning the way science is being taught in pre-schools. The interviews and conversations held with the teachers presented a picture showing no organized or compulsory curriculum for science teaching in kindergartens. The urban kindergarten's teacher expressly complained several times in different conversations about the absence of any such program. She did, however, favorably point out a series of five visits to a science center (this will be discussed in the final chapter of the book, dealing with out-of-school learning).

From what both teachers said, it appeared that the following is their interpretation to what they are expected to do in regards to science: To deal with science when the appropriate opportunities present themselves, by their own choosing, as a part of the annual program dealing with different topics, and to teach science using their own judgment. This concurs with the view expressed by the kibbutz kindergarten teacher who stated that: "Science teaching (is done) . . . through space . . . through the legs, through field trips," and in another occasion, "science integrates in everything even without naming it." As an example, she pointed out that the pursuit of science takes place through the learning of holidays, for example: The Jewish holiday of Lag Baomer. Lag Baomer is a traditional holiday in which groups of youths collect vast amounts of wood, which they later use to create a bonfire on the night of the holiday. The teacher pointed out that the flame and fire used in this holiday can be used as an opportunity to pursue science, so that there may be an integration of science in many daily opportunities. However, it was clear from the interview that for her, using terms such as "burning" and "fire," and talking about the burned woods, is a kind of such an implementation of science teaching. I do not see this as a kind of scientific activity that fully exploits the scientific aspects of the situation due to the lack of experimentation, concluding, creation of any hypothesis etc. In other words, there is partial dealing with scientific concepts; however, there is a lack of scientific thinking.

Both kindergarten teachers mentioned that a few years earlier they received a science kit for their kindergarten which contained some equipment such as lenses, mirrors, distance measuring tools, etc. They also both participated in an in-service course which aimed at preparing them for using the kit. However, they both complained about the equipment not being durable enough to withstand the conditions of the kindergarten which did in fact not survive. The urban kindergarten teacher claimed that the very perception of a science corner in the kindergarten isolates the incidents and distances them from the daily activities, which she saw as an additional shortcoming in integrating the kit into the kindergarten. This means that the teacher sees a gap between the way which she has been directed to teach science by the ministry of education, and the kit which was received from the same office. When asked as to the written material used to guide her and supply her with activities, she answered that written material does indeed exist and includes various tasks, but is not very lucid and is rarely used in kindergarten activities. This complaint was shared by the other teacher. In summary, the teachers of both kindergartens criticized the kit immensely. In addition, much criticism was voiced toward the nature of the instruction that teachers receive in scientific topics. In the urban kindergarten, there was mention of the varying policies of the ministry of education toward the existence or non-existence of instruction during a certain year. In the kibbutz kindergarten, the teacher had troubles fingering out this instruction in any way.

Teachers' Views Concerning the IE Method. In both kindergartens we saw a positive attitude toward the IE. This attitude was expressed in two ways: practical and verbal. The practical expression of a positive attitude toward the program was evident in the change of willingness of the urban kindergarten to receive this activity and to free the necessary time for it. With the beginning of cooperation between L. B. and the preschool teacher, the teachers seemed to be a bit suspicious and mostly hesitant toward the compatibility of the learning material to the group of pupils in her kindergarten. She found difficulty in freeing time for the activities and it seemed that she preferred different alternatives. As the IE program progressed, her willingness to free the required time grew. In the conversations L. B. had with the kindergarten teacher while coordinating the continuation schedule by phone or during the visit at the kindergarten, the teacher expressed a positive attitude toward IE which, in my opinion is reason for the supposed change. The kibbutz kindergarten teacher showed willingness to participate in the program from the start. It was apparent that her satisfaction in the progress of the conveyance of the program led to the continuation of this willingness.

The urban kindergarten teacher commended significant aspects: the program's spiral nature allows broadening, deepening and repeating of matter studied earlier: the way that the matter is structured — first creating a basis of motivation which can then be built on: a goal exists, however the path leading to it is no less important and many new things are learned along it. Another advantage that was especially noticeable with the urban kindergarten population is that the things there are simple and familiar in principle. The kindergarten teacher worded it as: "We didn't bring intimidating instruments." For the kibbutz kindergarten teacher the fact that the program provided tangible products like napkin dispensers, which the children could take home, or a file where their "personal cards" could be filed, were fundamental advantages. She repeatedly mentioned the possibility of the children taking the napkin dispensers they had made, as a [Shavuot] holiday gift. She seemed to be pleased with this option. After the first session she filed the "personal card," while mentioning the advantage of having a file like that, which can demonstrate the accomplishments in the kindergarten.

An obvious drawback of the program is holding discussions in large groups. The supervisor mentioned this drawback after watching a session in the kibbutz kindergarten. The urban kindergarten teacher also suggested that dividing the group into 2 groups of 13 children would also be an improvement. The kibbutz kindergarten teacher said that work in small groups, as suggested by the supervisor, demands the wearying and tiring matter of repeating the same activities many times. In this matter, she said in her interview: "I think that even in a small group, the remarkable students will stand out and the quiet ones will be quiet." She did not agree that the large group should be divided. This is possible, according to her, particularly because the group discussions are followed by small group hands-on activities where all of the children can express themselves. She also mentioned, with satisfaction, the contribution of the realizations that the activities invite: "children at kindergarten age need something real."

TEACHING STRATEGIES USED BY THE TEACHERS

The following three categories were identified in regards to the teachers' teaching strategies: strategies advancing scientific knowledge, strategies advancing scientific reasoning, and strategies used to recruit children's attention and advance coherent understanding of IE. Each of these categories was divided into several sub-categories which are detailed here.

Strategies Advancing Scientific Knowledge

Scientific concepts are a fundamental part of the suggested program. They are learned in association with the IE. Specifically, they are learned within its scientific aspect. The level of difficulty is supposed to be suitable for kindergarten children. Different central scientific concepts were discussed in the two IEs: in the "sending a parcel," the scientific concepts included *weight* and *material properties — water resistance, hardness, resistance to tear, stability etc*., and from the "guest visits the kindergarten," *solids – liquids, concentration,* and *stability* were included. These were not the only concepts. For instance, the concept of melting came up on both inquiry events.

The strategies used to teach the scientific knowledge were divided into the following sub-categories: (1) familiarization of new terms by announcing part of the term, (2) explaining the new term by referring to its verbal meaning, (3) reinforcing understanding by purposely referring to a wrong possibility, and (4) using analogies to reinforce understanding.

Familiarization of New Terms by Announcing Part of the Term. A typical strategy used by both teachers was to say part of a term and let the children complete it. Learning to pronounce and use the term is a large part of understanding it.

. . .

5. Teacher	How do we make raspberry juice?
6. Children (together):	With this pink thing.
7. Teacher	So what is it called?
8. Almog	(Trying to remember).
9. Teacher	Concen . . . . . .
10. Almog	Concentration.
11. Children (shouting together):	It's heavy, it's similar to black.
12. Teacher	What's this? (presents the bottle of concentrate)
13. Children	Concentrate

Explaining the New Term by Referring to its Verbal Meaning. The following paragraph is a continuation of the previous one:

14. Teacher	If I drink this (the concentrated liquid) what will I feel?
15. Teacher	Why will I feel such sweetness?
16. Child	Because you didn't add any water.
17. Teacher	Because they took the raspberry juice and concentrated it . . . It's very concentrated and that's why it's so sweet. Do you think it'll be tasty for me?

The teacher (lines 5 to 7) determines that the children are not quite familiar with the term *concentration* and mediates toward familiarity with it by adding part of the term (line 9). After one child completes the term, the teacher goes on to check whether the other children can now use the term, and by pointing to the bottle, i.e. use of a materialization (line 12), she makes sure that the children are familiar with the term and know how to use it (line 13). With the children now familiar with the term, she moves on to mediate their real understanding of the term. She doesn't define it, but rather goes back to the taste of the concentrated liquid (line 14). The children can now imagine its taste and connect that to the concept of concentration, as indeed one of the children explained that no water was added to the concentrated liquid, and hence it is sweet (line 16). The teacher continues and explains why the liquid is sweet by using the new term in different grammatical forms: as a verb and as an adjective (line 17).

Reinforcing Understanding by Referring to a Wrong Possibility Purposely. Purposely referring to an incorrect possibility was a common strategy used by both teachers. Session 6 of the "sending the parcel" IE takes place after the children come back from visiting the post office. Discussion of the balance scales is held without the scales themselves (after using the scales in the previous sessions). Measurement of the weights with the scales is not easy for the children. They need to understand:

a) The purpose of the weights.
b) The significance of the need to balance the two sides for measurement to take place.

Here is a segment from the discussion:

1. Teacher — How many holders does the scale have?
2. Children — Two

At this point, the questions were familiar, presenting the topic the teacher wanted to pursue. It is worthwhile to realize that she could do so by mentioning that the scale has two holders and then continue to what she wanted to discuss. However, she preferred to use questions, presumably because in such a manner, the children who feel they know, might be more motivated and willing for the next challenge.

. . .

7. Teacher — True, we check using balance scales with two holders, one with
8. what we want to weigh, right? We put weights on the second
9. holder and . . . when will the scale tell us the weight? (No response from the children and a silence is heard for a few seconds).
10. Teacher — If I put the parcel on one holder and I put a weight which
11. doesn't manage to bring down the parcel on the other holder, is
12. that the weight of the parcel? (The intention is to the weight of
13. the weights)
14. Child — Yes (quiet around him)
15. Teacher — That's it? That's the weight of our parcel?
16. Hen — It means that the parcel is heavier.

Line 7 presents a difficult challenge for the kids. The teacher, who realizes that the children are not responding, goes on to clarify her question. She provides a wrong assumption — lines 10–12, which results in one wrong answer — line 14. This wrong answer does not get an active response. Rather the teacher waits for a few seconds (there was a silence for few seconds). Then the teacher repeats her question implying that they needed to think further: that the previous direction was not the one she expected. She elicits a correct conclusion from one of the children (line 16). The teacher continues:

17. Teacher It means that the parcel is heavier. When did we reach that
18. conclusion? When did we get the result of the weight of the
19. parcel? How did we reach that result? When one holder is up
20. and the other is down?
21. Children (enthusiastically) When they are both the same.
22. Teacher When we reach a balance. When we kept adding until they
23. were balanced.
24. Teacher Yes, that's how we weigh with a scale

In line 17 the teacher repeats the correct answer. She now moves on to her original question in lines 17–19 in different forms. This is done to make the point clear. In lines 19–20 she again uses the wrong possibilty method. Her response invites the children to reject her proposal and look for another one. In line 21 few children reach the right conclusion and the teacher reinforces their answer.

It must be noted that this discussion took place without any real exhibits. To clarify the weighing process without the presence of the scales, the teacher mediated through *question asking, suggesting incorrect answers, reinforcing correct answers* and repeating them.

Using Analogies to Reinforce Understanding. In the summarizing session (observation 6) of the sending a parcel IE the teacher checked whether the children understood that heavier objects contain less air than lighter ones:

1. Teacher What materials did we check with the scale, which are similar to crisps, that we took apart and saw that it had air?
2. Christina Foamed plastic
3. Teacher . . . and what materials are more similar to gumdrops?
4. Child Gravel and also nails.

In this example the teacher seems to make use of analogies. She asks the children to suggest materials that are similar to crisps (a snack whose structure resembles foamed plastic) and others that are similar to gumdrops (which have more similarity to a stone).

Strategies Advancing Scientific Reasoning

Scientific reasoning is defined here as the skills which enable one to observe, hypothesize, use appropriate apparatus, measure, interpret data, and draw generalizations. Teaching strategies connected with these skills were identified: directing to specific features in observations, advancing causal thinking and thinking in a multi-constraint environment, and encouraging generalizing.

Directing to Specific Features in Observations. Observations have an important role in science and require one developing appropriate skills. For example, children should be able to focus and concentrate on relevant parts of the objects, things, or processes which are relevant to the study at hand. In the following episodes, we describe how the teacher directs the children to conduct observations. In the sending a parcel IE program, after the children learn about the weight differences between the different materials that were tested, a question about the reasons for these differences is asked. To do this, the teacher brings a small stone and some foamed plastic. The teacher asks that they pass these around themselves and says:

5. Teacher Look, it's bigger than the stone. How can that be? (referring to the weight of the piece of foamed plastic)

The children are not explicitly asked as to the reasons but are *directed toward raising hypotheses*, based on the observation, which would explain the foamed plastic being bigger yet lighter.

6. Hen I know, the size doesn't matter. How light it is is what counts.
7. Matan If you step on a stone it doesn't break. It's easier to take apart.
8. Hen Because foamed plastic is weak and stones aren't weak.
9. Teacher So what is it?
10. Hen It's strong.

It is important to mention that the teacher does not respond to answers that she was not expecting. She asks to continue passing the stone and foamed plastic around and asks again:

11. Teacher Why do you think the stone is heavier?

And some additional hypotheses follow here:

12. Ariel It's weaker.
13. Matan It's a kind of metal.
14. Child They're both light . . . they're just hard to break.

It appears that at this stage the children did not go in the direction that the teachers desired at all. They are focused on the idea that the foamed plastic is weak, probably because they feel that they can easily break it, as opposed to the stone which requires a much greater force to break. Therefore, she needs to clarify her question by another reference to the main topic:

15. Teacher But I purposely wanted to give you a small piece. You can still notice that the small stone is heavier, even with these tiny pieces.

And in response the pupils make additional hypotheses:

16. Child Size is not important.
17. Hen Weight is what gives us the strength.
18. Christina When its material is strong.
19. Child A stone is a stone.

As can be seen from the previous paragraph, the teacher still did not get the responses she was expecting. Now she focuses her questions and directs them to "feel" what makes the foamed plastic lighter than the stone. This is still hard for the children.

20. Teacher What does foamed plastic have that makes it lighter? Do you feel what it has that makes it lighter?

She then continues in the same direction while raising hypotheses:

21. Hen You can crumble it
22. Teacher Look, do you see the holes? What's inside?
23. Hen What's inside?
24. Teacher If we have a material that has holes inside, what does that mean?
25. Child Sponge.

26. Teacher What do holes have inside of them?
27. Child Nothing
28. Hen Air
29. Teacher Eh . . . Foamed plastic has lots of air in it, so now do you understand why it is lighter than the stone?
30. Child Because it has air in it

After revising her question (line 20) she still did not get the answers she was expecting (line 21). Only now does she directly ask the children to look at the holes (line 22). Again, it is important to note that she does not actively respond to answers that are wrong from her perspective, and after one child answered that there is nothing in the holes the room is filled with silence. This strategy was typically used by both teachers on different occasions. The children, who probably understand that silence means that they should think more, indeed suggest the answer that the teacher was looking for (line 28) and which she reinforced (line 29).

Here is another example on the same topic, but on a different occasion:

During the revision that took place in the fourth session, one of the children suggested sending crisps. The teacher shows enthusiasm toward this idea and asks the assistant to bring a packet of crisps, with which they then conduct the observation:

7. Teacher Shula is passing the crisps. You must all take one crisp, but before you eat it, you must check to see if there are air bubbles.
8. Children There are. There are.
9. Teacher So what is there?
10. Children Holes.

In this example the teacher *shows flexibility, accepting one of the children's ideas and pursuing it further*. In this case the teacher directs the children to look for the bubbles in their observation.

Advancing Causal Thinking and Thinking in a Multi-Constraint Environment. Research, in essence, is based on finding relationships between variables. As can be seen in the previous examples, an effort was made to find a connection between the weight of an object and its structure — does it have holes? We found that the IE program provided the teacher with many situations where she could advance the children's skills to find relationships between variables. Here are some illustrative examples:

Teacher Should we send chocolate?
Matan No
Child Because it will melt . . .

And afterwards:

Teacher When does the chocolate melt? When what happens?
Matan Melts faster when it is heated
Child From the heat

In this example, the children were asked to connect between heat and melting and to realize that because of this, a chocolate should not be included in the objects that will be sent in the parcel. In addition, the program invited opportunities where the children needed to take into account different considerations, and realize that although a solution may fit to fill one constraint it may fail to fit another. They were also exposed to the need for compromise. For example:

1.	Teacher	What would it be better to send a parcel in, what did we decide?
2.	Hen	Cardboard
3.	Teacher	Why did we decide on cardboard?
4.	David	Because it has space for crisps
5.	Teacher	Right, what about the weight?
6.	David	You can also put it in metal
7.	Teacher	Let's say that it would go in (the crisps), should we take a metal case?
8.	Child	No, because it's heavier
9.	Teacher	What would happen if it was heavier?
10.	Child	We'd pay more money

Simple as it may seem, it is not so simple for the children. The children need to understand that both metal and cardboard boxes are good materials for a parcel container. In both there is space for the crisps. They must understand that, while it is possible to use a metal container to move the goods, and it may even be preferable by some aspects, like withstanding higher loads or resistance to tearing, they must still take weight and cost into consideration. If the parcel is heavier they will need to pay more, as they were told in the post office, when the children visited it.

Moreover, as we shall see, the IE invited the teacher, on occasion to search for reasons as to the connections between and among variables. It is important to have skills in finding connections between variables through observation and measurements. This skill may even be considered a great achievement in developing the child's scientific thinking at this early stage of life. With this in mind, in many cases the program invited the teacher to encourage the children to give explanations on the nature of these connections. Finding a connection between variables does not necessarily indicate understanding the nature of the connection. For example, why is a material that has many holes lighter? In another example, after reaching an agreement that chocolate is included in the group of items unsuitable for sending in a parcel, the teacher sees fit to elaborate on the cause of melting. She conducts an experiment where the heating mechanism is a candle and the margarine in the pot hardly changes in the few minutes that the heating takes place. The teacher draws the children's attention to the "problem":

Teacher Now look at what happens here. It happens very slowly. Why? Does anyone have any idea?

In response, the pupils raise ideas about the causes. In this case the cause was the fact that there was only one candle. Pursuit of the cause was the teacher's idea and was an elaboration on the suggested IE frame.

Encouraging Drawing of Generalizations. One important thing in science is drawing generalizations. After all, we do not want the child to refer the findings of an experiment only to that specific experiment, but rather to be able to understand that the findings obtained in one experiment may be generalized. The IE program enabled many such situations. For instance, after categorizing flowers as unsuitable for sending in a parcel because they wither, a generalization is made about all plants. The following are another two detailed examples of this issue:

Example 1

After concluding that a lighter object has holes which contain air, the teacher directs for generalization —

1. Teacher What can we conclude from our observation?
2. Avichai Whatever has no air is harder
3. Teacher And what about its weight?
4. Avichai It is heavier
5. Teacher Avichai has reached a very important conclusion, that whatever has air inside is heavier?
6. Avichai No, lighter

Here one can see the use of a sort of scientific language by the teacher (line 1) — conclusion based on evidence, i.e. the observations. It is interesting to note that one of the children was immediately able to formulate a generalization (line 2), although not the one the teacher desired. However, after directing the children (line 3), the same child revised his answer (line 4). Now, the conclusion does not refer to a specific object, but is rather a kind of principle — the children have reached a generalization. To confirm that the children did indeed understand the principle, she uses a known technique discussed previously — providing a wrong possibility (line 5).

Example 2

This example is from the activity where the children built a wooden napkin dispenser. Its base could be wide or narrow. At the request of the teacher that the napkin dispenser be stable, one of the children suggested the wider base option. The teacher refers to the child's suggestion:

Teacher Raz used a wide base. Who has any conclusions from what Raz is doing?

This was hard for the children and they didn't respond. The teacher used a familiar strategy for the kids: she began a sentence and asked them to continue:

Teacher The conclusion starts from the following sentence, listen to my Sentence: the wider the basis is . . . the wider it is . . .
Raz The more stable it is.
Teacher The napkin dispenser is more stable.

It is important to mention that this generalization was not completely absorbed by the children. In another opportunity the teacher repeats the same idea and continues a sentence:

Teacher The wider the napkin would be, it would be . . .
Child Taller.

Or in the concluding meeting when the second researcher (H.E) asked the children about the differences between the two possibilities of building the napkin dispenser with a wide or a narrow basis, no one mentioned its stability, even though they did provide good answers:

Researcher	How is it better to build the napkin dispenser, like this (points to the narrow base) or like this (points to the wide basis)?
Child	A wider is better.
Researcher	Why?
Child	Because one can put more napkins.
Raz	In here (points to the suspender with the narrow basis) I can't even put my hand inside.

Strategies Used to Recruit Children's Attention and Advancing Coherent Understanding of the IE. One major advantage of the IE is the presence of the IE's main problem which enables the teacher:

a) To quickly recruit the children's attention, and motivate them for the upcoming activity.
b) To create connections among different parts of the IE and thus lead to a coherent understanding of the IE.

Brief Reminder. For example, through a short question, why did we think about the parcel? At the beginning of the second session of the "sending a parcel" IE, the teacher moved quickly to the idea of weight. It seemed very much connected to the parcel IE's problem, and no long introduction was needed to get into the new topic.

In the other IE, the teacher could easily move to the topics of measuring the table for the tablecloth, making the raspberry juice, or melting chocolate or margarine to prepare snacks because of the direct relationship to the main problem of the visitor to the kindergarten. This made the transitions feel very natural.

Encouraging Meta-Cognition. The presence of the IE's main problem helps the teacher to ask questions which force the children to remember the question they were dealing with and the process they used to reach conclusions. For instance, how many jars did we decide to make for when the visitor comes? Why did we decide this? How did we prepare the chocolate fondue? How did we come to know that the chocolate melts when it is heated? Why didn't we choose the metal parcel? How did we come to know that a lighter object has more air? How did we know how much raspberry concentrate to put in the jar? How did we know that the heavier the parcel will be, the more expensive it will be to send it? How did we measure the parcel's weight?

DISCUSSION

The current research examined the effects of the Inquiry Event teaching method. The IE was developed to give the kindergarten teacher a comfortable environment for science teaching. As with many other scientific curricula, the IE is also based on pedagogic approaches such as: inquiry learning, problem based learning and authentic learning. So, one can ask, what is the difference? For me, while there are some similarities between the IE and the other curricula, there is also a significant difference. The IE is built in a way that considers the pre-school teacher's needs first. The teacher's needs are expressed in the following manner: (1) the inquiry events are situations that are familiar in the teacher's everyday life in and out of the kindergarten. (2) The scientific aspect of the inquiry event is only part of the whole event. In addition, it should be noted that the scientific aspect is systematically built-in to the curriculum. The assumption is that if the kindergarten teacher deals with everyday situations, where most of the aspects deal with problems which are familiar, and the scientific aspect has organized guidance, it will be easy and natural for the teacher to acquire the missing scientific knowledge and mediate it efficiently to the kindergarten children. The present study tested this assumption. The findings do indeed verify this assumption. To understand how the IE indeed influenced the science teaching in the two kindergartens in which it was tried, let us first refer to the situation of science teaching in those kindergartens before the IE curriculum was executed.

Our findings reveal a gap between two contradicting lines of evidences. According to one line of evidence, there were little dealings with science prior to the IE program. The two teachers complained about the absence of any structured curriculum, and about the science kit they got a few years earlier as being unsuitable for children. Moreover, from the interviews with the urban teacher, the more meaningful scientific activity were the visits to a science center, outside the walls of the kindergarten, which were not lead by the kindergarten teacher, but rather by the science center's staff. According to the other line of evidence is the picture according to which science teaching is integrated into "everything." So, on one hand it seems that the teachers are not satisfied with the absence of a systematic scientific curriculum and find it difficult to mention specific episodes of science teaching, and on the other hand, it seems that science teaching is done everywhere, anytime, "through our feet, through field trips and all the subjects that we're working on." How then, does the gap between this discontentment and the report, according to which, science is seemly integrated in "everything we do," develop? To bridge this gap we must first understand the central approach, according to which most kindergarten teachers in the country (including those in the current study) have been trained. This approach is the integrated approach, according to which the daily activities should be approached from a variety of different aspects, because even the most banal topic has great educational and research potential. Even though the teachers understand the central idea behind this approach, from the interviews with them regarding the state of science teaching in the kindergarten prior to the study, we feel that implementing this approach is meaningless. Teachers do not understand and do not have the necessary tools for implementing this integrated teaching approach. After all, for kindergarten teachers who do not posses good scientific background it would be hard to "find the science" in the daily situations which they are confronted with. I warn that such amorphous approaches according to which one can teach science "everywhere," "anytime," and with no systematic curriculum, might lead, especially in the case of preschool teachers, to the situation, as was the case in the two kindergartens in our research, of an illusion of science teaching with no real teaching. Surely enough, lack of significant dealings with science often arises when trying to "combine" science with everything (Schoeneberger and Russel, 1986).

The IE was found to be an efficient learning method. The findings show that the IE helped teachers to advance both scientific knowledge and scientific reasoning. As explained in the first chapter of the book, the term 'science' is used to describe both a body of knowledge, i.e., the variety of scientific concepts as well as the activities that give rise to that knowledge, i.e., observing, asking questions, hypothesizing, using appropriate apparatus, measuring, recording data, interpreting data, and formulating theories or models. Advancing scientific knowledge in kindergarten children means that first they should be familiar with the terms verbally. We found that both teachers, in many cases, familiarized the terms by starting to say part of the term and letting the children complete it. Also they explained the meaning of the term verbally and used different grammatical forms so that the children could internalize the concept. The fact that the concepts were familiar to the teacher made it fairly easy for them to teach the scientific concepts in the same way that they teach language.Thus, they could use analogies to confirm the children's understanding. In advancing scientific reasoning processes, the teacher also encouraged children to draw generalizations. This is indeed a very important result. Being able to draw generalization is considered as a higher order thinking skill (Zohar, 1999). If one purpose of science education stated in the first chapter, is its ability to develop children's cognitive capabilities, this finding shows that the IE does indeed address this issue. Moreover, we found that the IE enabled the two preschool teachers to expose children to real life situations where they needed to consider several conditions at the same time. Sending the parcel in a nylon bag will be cheap because it is light and also because it protects the parcel content from getting wet, but at the same time it is not strong enough and therefore a cardboard box will be better, even though it does not possess two of the nylon bag's characteristics. This demands *multi-consideration thinking*, which also is a type of high-order thinking, that should be nurtured in children.

One interesting result is the existence of the main IE problem. It was found that this helped the teacher present connections between the different parts of the IE. It was as a kind of a powerful background that allowed the teacher to deal with a variety of activities. In this way it was easy for the teacher to recruit the children's attention to new concepts and tasks, as soon as they were convinced of the connection between the new activity and the main IE's problem. In addition, the IE's main problem also served as a kind of glue that enabled the connection of the different activities to a coherent story. My interpretation is that these "stories" made the science activities more relevant, both to the teachers and to the children. This relevancy made science easier for the teachers to teach and for the children to learn. Also, we believe that another advantage was the fact that the IE presented a real problem in a sense that all the activities aimed at achieving a concrete goal. In chapter 3 it was argued that children tend to employ engineering models of inquiry in which they explore the reasons for achieving a desired effect, rather than scientific models. In the same manner it might be that the practical nature of the IE, i.e., that the children act in order to execute some real objective like having the visitor or sending the parcel, is an advantage because it fits the natural way children learn.

In summary, although this study took place in only two kindergartens and consisted of only two inquiry events, it seems to lead in the direction of building a curriculum which considers and emphasizes the teachers' needs and not only those of the children. This may lead to more efficient K-2 science teaching. It is particularly important at the K-2 level which the literature sees as one of the weakest links of science education. There is room, of course, to broaden the IE to wider populations and additional *inquiry events*.

CHAPTER 5

BRIDGING IN-SCHOOL AND OUT-OF-SCHOOL LEARNING: FORMAL, NON-FORMAL, AND INFORMAL

To understand fully children's science learning, one should look not only at learning that takes place in the kindergarten and primary school, but should also at the learning that takes place out-of-school. This is very important considering the fact that 85 percent of the time children are awake is spent outside of the classroom (Medrich *et al.*, 1982). Children's life experiences, both in and out of school have profound effects on their achievements in school and their functioning in society (Resnik, 1987). Support of the importance of informal experiences can be found in the National Science Education Standards (National Research Council, 1996), which state that museums and science centers "can contribute greatly to the understanding of science and encourage students to further their interests outside of school" (p. 45). Museums and science centers are just examples of out-of-school learning and one may broaden this idea to other forms of learning of this type. Gardner (1991), goes even further to argue that whereas schools have become increasingly anachronistic, museums have retained "the potential to engage students, to teach them, to stimulate their understanding, and most important, to help them assume responsibility for their own future learning" (p. 202). Indeed, Stevenson (1994) reports that as opposed to a normal museum visit where visitors typically display fatigue after 30 minutes, Launch Pad science museum visitors usually displayed little or no reduction in concentration even after 60 minutes.

Before moving on to describe and discuss the advantages of out-of-school learning, it is also important to consider the critiques. In responding to Stevenson's findings, Rennie and McClafferty (1996) raise the following questions: are visitors concentrating because they are learning the scienctific concepts that are portrayed by the interactive exhibits or are they just having fun? In searching for answers to such questions some researchers "used the term 'edutainment' to describe science centers, politely suggesting that perhaps the entertainment dimension is more successful than the educational one" (Rennie and McClafferty, 1996, p. 55). Shortland (1987) and Wymer (1991) suggested that education loses out when entertainment become a major consideration. Shortland bluntly said that "When education and entertainment are brought together under the same roof, education will be the looser" (p. 213). Ansbacher (1998) argues in this regard that placing emphasis on museum learning being fun may be antithetical to the learning outcomes desired by teachers. Citing Dewey, he reasoned that if the experience is mainly fun, the learner may have learned something, but not necessarily what the teacher or museum educators had planned. The implication is that visitors may learn to pursue further fun rather than further learning. The 'fun-emphasis' problem was already mentioned by Champange (1975), a quarter of a century ago, when he described a six hour visit to a science museum as entertaining, but unfulfilling. Champangne raises the following 4 reasons to his dissatisfaction:

1. the real meaning was obscured,
2. some of the demonstrations involved 'sloppy science',
3. science and technology were presented as ethics-free, and
4. science was dishonestly presented as easy and unproblematic.

The last reason is very serious because firstly, it obscures what science is really about: the asking and answering of questions about how the world works, and secondly, such presentation suggests that scientists are very smart and possess superhuman intellectual capacities, which enable them to accomplish anything — just point them to the target and in a short time they will get there. This may also insinuate that science is not for all students.

Parkyn (1993) also argues that "scientific phenomena are presented not within a conceptual framework but as an endless series of unconnected, entertaining magical events" (p. 31). These criticisms do not seem to have been refuted: in fact they have been reiterated (Rennie and Williams, 2002).

With that said, one should bear in mind that in spite of the critiques, most science centers do believe that the visits to the center enhance visitors' understanding, or at least awareness of science. Indeed, Falk and Storksdieck (2005), based on past and present literature, claim that whereas only a few years ago it could be briefly stated that it was unclear whether visitors to museums truly learned, that is not the case today. However, Griffin and Symington (1997) found that unfortunately, teachers who themselves planned scientific fieldtrips to science centers, displayed little recognition of the different learning environments or learning opportunities that museums present. Furthermore, the authors found that teachers may not necessarily have explicit goals for the visit, and are unable to connect the experience to the classroom curriculum. They suggest that teachers might feel intimated when they take classes to museums. They also have many management concerns: losing children; risking the reputation of their school; not knowing where to go; and being asked questions which they cannot answer. These are probably some of the reasons why students participating in teacher-led school fieldtrips, in many cases, are not aware of any specific goals that these visits may hold and thus may subsequently be unprepared for learning (Griffin and Symington, 1997; Orion and Hofstein, 1994; Stroksdieck, 2001).

This chapter, which thoroughly examines the idea of out-of-school learning, aims at providing educators, especially those who work in K-2, with an insight to the topic both theoretically and practically, so that they will be able to fully exploit the potential that field trips may offer. In the following I will discuss the difficulty in defining out-of-school learning and then I will raise the question of whether we should deal with out-of school learning in the *in-school* systems.

WHAT IS INFORMAL LEARNING? TOWARD A DEFINITION

There is some sense that in-school learning is formal learning and out of school is informal *tout court*. For example, Gerber, Marek, and Cavallo (2001) argue that,

> in essence, the informal learning can be defined as the sum of activities that comprise the time individuals are not in the formal classroom in the presence of a teacher. (p. 570)

Resnick (1987) differentiates sharply between the nature of "school learning" and "other learning." Based on the literature Gerber *et al.* (2001) argue that while formal learning environments are characterized by their highly structured nature, the informal learning environments are less structured, and managing the learning is shifted from the teachers to the students. I do not agree with such a comparison. Let's consider for instance a field trip to a science museum. First, it is outside the classroom, so learning in the museum is, according to the above definition, indeed, informal learning. Indeed, the children may more than likely be invited to free, unguided visits, in which they may approach different exhibits as they desire. Yet, in many cases, part of the museum field trip includes a highly structured visit. The children may conduct experiments, fill pre-prepared work files and follow a guide. I agree with Dierking (1991), that such sharp distinctions between formal and informal learning are inappropriate, as he sees the physical setting as only one of a number of factors governing learning. According to the author, "learning is learning, and it is strongly influenced by setting, social interaction, and individual beliefs, knowledge, and attitudes" (p. 4). Gilbert and Priest (1997), argue that "if teachers in school and adult companions during museum visits both see themselves promoting meaningful activity by means of focused conversation, then it does seem very likely that the learning taking place would be similar in type and quality" (p. 750).

The problem of distinguishing between formal and informal learning may also be found in Hofstein and Rosenfeld (1996) who argue that,

> There is no clear agreement in the literature regarding the definition of informal science learning . . . . The major difficulty in defining informal science learning is determining whether or not informal science learning can take place within formal settings. In other words, does the term have distinct, clear-cut attributes of its own (in which case it may occur in formal as well as informal settings) or must this term be understood as necessarily contrasted with formal learning (in which case it cannot occur in formal settings)? (pp. 88–89)

A better distinction, in my opinion, is one that takes into account not only physical differences, i.e. in or out of school, but rather includes other factors as well, such as motivation, interest, social context and assessment to distinguish between three types of learning: formal, informal, and non-formal (Maarschalk, 1988, in Tamir, 1990).

Non-Formal Learning

Non-formal learning occurs in a planned but highly adaptable manner in institutions, organizations, and situations beyond the spheres of formal or informal education.

It shares the characteristic of being mediated with formal education, but the motivation for learning may be wholly intrinsic to the learner.

Informal Learning

Informal learning applies to situations in life that come about spontaneously; for example, within the family circle, the neighborhood, and so on. These are reflected in what a person is reading, viewing and listening to, and also in his or her hobbies and social life (Maarschalk, 1988, in Tamir, 1990, p. 34). Informal learning is distinguished from the other two by having no authority figure or mediator. The learner is motivated intrinsically (Csikszentmihalyi and Hermanson, 1995) and determines the path taken to acquire the desired knowledge, skill, or abilities. Table 1 summarizes some of the differences among these three types of learning. Dividing of out-of-school learning into informal and non-formal categories help to achieve a better understanding of the characteristics of out-of-school learning. Yet, a variety of institutions are still hard to categorize as non-formal, because they are still different despite the fact that their activities might share some similarities. One striking difference concerns the degree by which one may manipulate the exhibits.

As opposed to other non-formal locations, museums and scientific centers include, to a large extent, interactive science exhibits. Rennie and McClaffery (1996) distinguish between 'interactive' and 'hands-on' exhibits. According to the authors, hands-on exhibits require the visitor to have some physical involvement with the exhibit. However, while hands-on exhibits are passive, interactive exhibits are active and respond to the visitor's actions. Consider for example, a visit to the planetarium. Here, the visitor usually enters a room and the explainer, by pressing different buttons, displays the star system. The explainer shows different patterns in the sky for example, the Ursa Major, by lighting up different areas of the room's ceiling, which

TABLE 1. Differences between Formal, Nonformal and Informal Learning

Formal	**Non-Formal**	**Informal**
Usually at school	At institution out of school	Everywhere
May be repressive	Usually supportive	Supportive
Structured	Structured	Unstructured
Usually prearranged	Usually prearranged	Spontaneous
Motivation is typically more extrinsic	Motivation may be extrinsic but it is typically more intrinsic	Motivation is mainly intrinsic
Compulsory	Usually voluntary	Voluntary
Teacher-led	May be guide or teacher-led	Usually learner-led
Learning is evaluated	Learning is usually not evaluated	Learning is not evaluated
Sequential	Typically non-sequential	Non-sequential

represents the sky. The visitors themselves are not active and the exhibit does not respond to any of the visitors' actions. Another example is a visit to the zoo or to the aquarium. Here again, no one expects an animal to respond to a visitor's action, which may, at times, be forbidden. Yes, the visitor may, sometimes, feed the animal or touch its fur; but usually it is the animal that decides how it wants to respond, if at all, to the visitor. This contrasts with museum or science center exhibits which are usually designed to be interactive: for instance, the visitor may interact with a model representing an airplane. He or she may change the angle of one the airplane's model wings and as a result, the airplane might change its position.

By responding to the visitors' actions, interactive exhibits invite more actions from the visitors and provoke further interactions, and a kind of *man-machine dialogue* is developed. According to Rennie and McClaffery (1996), an important difference between hands-on and interactive exhibits is that hands-on does not necessarily mean 'mind-on'. The authors cite Lucas (1983), who pointed out that, "It is false to assume that any physical manipulation of an exhibit provokes intellectual engagement" (p. 9).

Borun and Dritsas (1997), identified seven exhibit characteristics that attract and hold the attention of family groups. These describe the desired characteristics of interactive exhibits that are usually placed in science centers, not in nature, parks or zoos. The characteristics are:

- Multisided: the family can cluster around the exhibit.
- Multiuser: interaction is allowed for several sets of hands (or bodies).
- Accessible: comfortably used by children and adults.
- Multioutcome: observations and outcomes are sufficiently complex to foster a group discussion.
- Multimodal: appeals to different learning styles and levels of knowledge.
- Readable: text is arranged in easily understood segments.
- Relevant: provides cognitive links to visitors' existing knowledge and experience.

In summary, the terms out-of-school learning and informal learning in the literature are usually interchangeable. I argued that defining informal learning as learning which occurs out of school is too simplistic. A better distinction, which captures the characteristics of out-of-school learning, is between informal and non-formal learning. I also claimed that we can distinguish between two institutions where non-formal learning takes place: those that possess hands-on exhibits and those that include interactive exhibits as well. Another distinction which might provide insight as to the nature of out-of-school learning is based on the frequency to which we attend the place where the learning occurs. In my view, since informal learning occurs spontaneously, it is more likely to occur in places within our day-to-day routine, such as homes, yards, parks or streets, and even at school — especially at break times. Since we only visit places such as museums, zoos, planetariums, or aquariums occasionally, it is more likely that non-formal learning will happen there — it is more likely that these visits are prepared to some extent. We also tend to participate in structured activities in those institutions, especially if the visit is in the framework of a school scientific fieldtrip. Fig. 1 summarizes the differences described above.

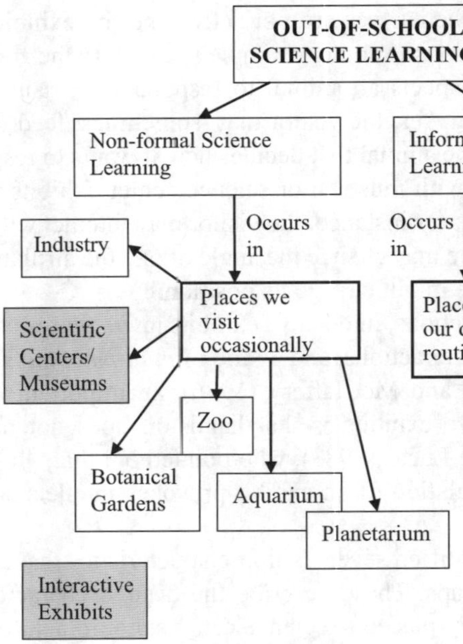

Figure 1. Informal and Non-Formal learning.

The title of the current chapter, *Bridging in-school and out-of-school learning: Formal, Non-formal and Informal*, implies that we should bridge between out-of-school and in-school learning. In the following section I would like to take a step back and ask, should we indeed, bridge the two?

WHY HAVE OUT OF SCHOOL ACTIVITIES DURING SCHOOL TIME?

As stated earlier, whether we plan it or not, informal learning occurs everywhere and all the time. We cannot avoid it. In addition, visits to museums, aquariums, zoos, etc., have become part of our way of life; so, the questions here are: if we experience informal learning anyway, why put effort into doing so during school time? Isn't it a waste of money? Wouldn't it be a waste of precious school time? I also mentioned that teachers have real difficulties when planning and carrying out scientific field-trips. Some of these difficulties stem from their lack of knowledge about organizing and conducting science field trips. These questions might be illustrated by results of a recent study that evaluated docent-led guided school tours at the museum of natural history (Cox-Petersen *et al.*, 2003). The study included observing about 30 visiting school groups in Grades 2–8. Some of their findings show that:

1. Tours focused on facts or stories rather than extensive ideas or concepts.
2. The scientific and historical vocabularies used during the tours were often too advanced for students.

3. Sensitivity to individual and cultural differences was rarely observed.
4. Closed and/or factual questions that did not require complex responses from students were observed. Questions were asked without follow-up, elaboration, or probing.
5. The structure and content of the tour provided minimal connections between the content of exhibit halls and the lives and prior knowledge of the students. Docents seldom provided analogies, information, or examples related to students' life outside the museum.

Moreover, in her recent review article on school group visits to museums, *Research on Students and Museums: Looking More Closely at the Students in School Groups*, Griffin (2004) concludes that,

> in general, school students still look, act, and are treated differently from children in family groups in museums (Hein, 1998, in Griffin, 2004). Their personal relationships within the group are limited, different expectations and constraints are placed upon them, and personal controls over their own movement, rest, and learning styles are often minimized. The school group is generally referred to and largely treated as a single entity rather than a group of individuals and the group's characteristics and needs are considered over the characteristics and needs of the individuals. (p. s67)

Considering Cox-Petersen *et al.*'s report as well as Griffin's conclusion, one may answer negatively the question of whether schools should also partake in out-of-school activities. Yet Griffin (2004) herself argues that with appropriate treatment, student learning can be facilitated. It seems as if there is a gap between the feeling that great potential may lie in school field trips, and some of the recent research results indicating that this potential is not fully achieved.

So far, I presented the voice of researchers or policy makers, such as those in national reports regarding their views toward informal learning. In most national reports it is advocated that informal learning should be pursued. Also, despite the gap mentioned earlier, most researchers would probably call for improving informal learning activities rather than give up and leave informal learning solely in the hands of families. To understand the reasons for this, it might be worthwhile to look for more theoretical explanations. But, before moving on to the theory, let me first present the voice of the teachers, the students, and the non-formal institutions staff.

Teachers' Perspective

Kisiel (2005) investigated the motivation that comprises teachers' agendas when leading student fieldtrips to science museums or similar sites. Eight motivations were identified. Included in the descriptions of these motivations, are the views of Ms. Meg Norton, a primary class teacher, who was a subject of investigation in Lucas's (2000) study; which aimed at describing the involvement of teachers and their students in a class visit to the science center.

1. Connect with the curriculum — teachers see fieldtrips as opportunities to reinforce or expand upon the classroom curriculum by providing an additional perspective, or a more meaningful connection, that can help them with part of the school curriculum. They also believe that students can gain knowledge, curriculum related or not, as a consequence of the visit. This is exactly what Lucas (2000) found

regarding Ms. Norton's agenda for the science center visit. According to the author, it appeared to be quite clear: she aimed to present the students in her class with many opportunities to learn about science and technology topics which they had already learned in class, and new topics which they hadn't.

2. Providing new learning experiences — teachers see the fieldtrips as opportunities to provide firsthand, rich, novel and entirely new learning experiences to students who may not otherwise have the opportunities. These experiences are believed to have a positive impact on student's development and future learning.
3. Providing a general learning experience — teachers see fieldtrips as opportunities to provide memorable learning experiences.
4. Fostering students' interest and motivation — teachers see fieldtrips as events that spark interest in some topics or concepts; hence, foster students' curiosity, motivation and will to discover more.
5. Providing a change of setting or routine — teachers see fieldtrips as opportunities to get out of the classroom and change the routine.
6. Promote lifelong learning — teachers see the fieldtrips as opportunities to show students that learning is possible beyond school, among friends and family. In this regard here is Ms. Norton's view,

> I have just tried to develop them personally into learners, and I think as a teacher that's probably the most important job I have to do: is to try and make them life long learners and to understand how they learn. (Meg, in Lucas, 2000, p. 531)

7. Providing students with enjoyment or reward — teachers recognize that the fieldtrip should be a positive and enjoyable experience for the students.
8. Satisfying school demands — teachers are expected to conduct fieldtrips, due to school policies or pressure from their colleagues.

According to Kisiel (2005), of all the above fieldtrip motivations, curriculum connection was the one most often mentioned most. However, teachers had different views about the nature of the connections. The following concepts of connections were identified by the author: *curriculum-related experiences* — students gain "hands-on" experience related to curriculum; *curriculum-related learning* — students gain content knowledge related to the curriculum; *connection to language skills* — students utilize language skills in an interesting real-world setting; *point-by-point connections* — students are directed to see how different aspects of the museum relate to different parts of the curriculum; *curriculum unit integration* — the museum experience is an integral part of a particular topic currently being studied in class, and the experience is directly related to current activities or projects; *curriculum unit introduction/review* — students are introduced to a curriculum topic which they have not yet begun in class, or they are reminded of a curriculum topic which they have already finished; *implicit/opportunistic connections* — students naturally relate their museum experience to their classroom experience. Teachers, if aware of these views about the variety of interpretations of *curriculum connections* are better able to decide what kind of connection they might seek for a specific visit, and plan the visit accordingly.

Students' Perspective

In reading the literature, I found scant evidence about students' perspectives on non-formal learning experiences. Usually, researchers are not interested in questions such as what the child thinks he will gain from his or her science field trip, or whether they think the field trip is important. Researchers are usually interested in and focus on children's attitudes and attitude changes toward science as a result of the fieldtrip, most often to a specific science center. An exception was Lucas's (2000) study. While investigating teachers' agendas for a class visit to a science center he also studied their students' anticipations. The author explored the students' perceptions of why they were being taken to the science center by their teacher and what they expected to do there. Here are some of the students' responses: Tom reasoned that his teacher was trying to make "science learning fun 'cause doing all the experiments and hands-on stuff like that is kind of different from just literature and writing it all down" (Tom, in Lucas, 2000, p. 532). Stuart said that "instead of writing it down and having to remember it, you go and test it out" (Stuart, in Lucas, 2000, p. 532). Bill said that boys "could learn more about science . . . in different ways" (Bill, in Lucas, 2000, p. 532). The researcher asked students who had already been to the science center what they thought they would do there with their classmates. Body stated that, "Umm, we'll split up into groups first and then we'll go around and, umm, if someone in the group doesn't understand how it works, we'll sort of, 'cause the theory sheets maybe too complicated for them, we'll explain it, explain to them what it does if we know ourselves" (Body, in Lucas, 2000, p. 532). Ian said, "just go round in groups and just explain to each other if we don't know, you know, if someone else knows, and just, you know help each other to understand it if they don't — 'cause we're going in groups — and just learn a lot more 'cause we're in groups than just with our family which, you know, you're always with your mum and dad telling the same things. But when you're going in groups you can learn a lot more" (Ian, in Lucas, 2000, p. 532).

In summary, the author concluded that students knew that they were expected to learn. They were equipped with a range of learning strategies, and they anticipated that the learning would be fun.

The above research has made me curious as to what my own children would say about science museums. I have 13 year old twins — a boy, Omry, and a girl, Shaked, and a 9 year old son, Ohad. I held a conversation with each one of them separately. Although they all like science, they do not want to become scientists. Omry prefers learning economics and Ohad wants to become a pilot. Shaked had the most original answer. She told me that, "science is *doing* and I'm a *being* person. I'm interested in more philosophical questions." At first thought it seemed that she missed the point. However, after deeper thought I can understand that this is probably her view because of how she has been educated. Science, to her, is indeed connected to doing. In museums she interacts with exhibits, at school she does experiments. In our conversation she repeatedly mentioned that science is related to facts. Facts that she probably perceived as acquired through doing. No one has yet emphasized the "*being*" part of science to her — that science does indeed deal with philosophical problems. This agrees with Champagne's (1975) critique of science museums, which I mentioned earlier, according to which science is dishonestly presented as easy and unproblematic; nothing to do with "being." One should not only relate this critique to science centers, but rather to the problem of formal science education as well. She might still not want to become a scientist after realizing that science is more than doing; however, she might change her views regarding science. Regarding museums, she said that she enjoys visiting museums since you "learn about the world; about phenomena. You learn how the world works." As for her friends, she said that she thinks that most of them "probably enjoy going to museums because it is a kind of change in the routine; a kind of a "day off." I don't think, though, that they like what is going on there. They do not understand science and do not really like science." Both, Omry and Shaked told me that there was never any connection between the curriculum and the science fieldtrips. The science teacher didn't even join their trips, but rather their class teachers who also organized and led the trips. Omry mentioned that he would prefer it if the science teacher would talk in class about what they saw in the museum. Ohad, the youngest, said that the most enjoyable thing to him is playing around with the different machines. He also mentioned that he enjoys building models in the museum, and that he loves going with his friends because they play together with the different machines and talk about them.

In summary, from this section it can be seen that children enjoy going on scientific fieldtrips. They are aware that they are expected to learn from the trip, and that it should not only be a "fun day", but rather a day where they enjoyably learn science.

Staff Perspective

Rennie and Williams (2002) interviewed, in the first stage of their research, a sample of 28 science center staff regarding their: understanding of science, where their ideas about science came from, what kind of image of science they thought the center portrayed, how it did this, and how successful it was. These included staff working in Administration, Education, Exhibit Design and Development, Visitor Services, and Explainers. The following are some of the main results relevant to this chapter:

1. Nearly half of the interviewed staff thought that part of the center's role was simply to display science and applications of science, with the aim of making people more aware of modern development, the history of science, and its role in modern day life.
2. Two thirds of the interviewed staff thought that part of the center's role was to influence the image that visitors held of science prior to their visit. They hoped that visitors would leave the center with a "more positive feeling about science," and would believe that science could be fun, interesting, easy to understand, and can benefit humans in their everyday life.
3. Over half of the interviewed staff mentioned that the center should provide visitors with the opportunity to gain more scientific knowledge, particularly through the interactive exhibits. Some staff members thought that it was important to recognize that people gain different understandings from exhibits, and that learning may not occur immediately, but rather the visitors' experiences may be expressed in the future.

It is interesting to mention that all but one of those who were asked whether they thought the center was successful in achieving its mission thought that it was successful to some degree.

In the second stage of Rennie and William's (2002) study, the researchers distributed a questionnaire to both the visitors and the science center staff members. Some interesting differences were found (the reader may need to read the original paper to gain a better understanding of this topic) between the staff members and the visitors. Staff members were more likely than visitors to respond that ordinary people can understand science and that the exhibits do not have enough explanations on science. In addition, there are some findings which may concern educators who plan visits to science centers. After the visit, visitors were more likely to respond that scientists always agree with each other, scientific explanations are absolute, science has the answers to all problems, and it is not likely that scientific knowledge will be misused.

From this discussion it appears that the reasons that teachers, science-center staff and children provide as to why scientific field trips to science centers are important, may be divided into two aspects: cognitive and affective. I will now focus on these two aspects of out-of-school learning.

A CLOSER LOOK: THE COGNITIVE AND AFFECTIVE ASPECTS OF NON-FORMAL LEARNING

A thorough comprehension of both non-formal and informal learning, must refer to the affective and cognitive axis of human behavior. I will now discuss how out-of-school learning impacts those two domains.

The Affective Domain

Scientific field trips to science centers can generate a sense of wonder, interest, enthusiasm, motivation, and eagerness to learn, which are much neglected in traditional formal school science (Pedretti, 2002; Ramey-Gassert *et al.*, 1994). Further, informal science centers provide opportunities for active science in non-evaluative and non-threatening environments that invite girls to take on the challenge of a subject that is traditionally viewed as male-dominated (Ramey-Gassert, 1996). Therefore, scientific fieldtrips may play a significant role in inculcating positive attitudes toward science among children, in boys and even more importantly, in girls. In this regard, Hodson and Freeman (1983) state that "the image of contemporary science and of scientists which is presented to young children (under 12) is . . . of great importance in forming their attitudes and determining their choices." Positive attitudes toward science as early as kindergarten and primary school are tremendously important as many latent scientists appear to make early decisions about their careers (Blatchford, 1992). This concurs with the finding of Musgrove and Batcock (1969), who found in their study that a third of 338 science and engineering students at the University of Bradford, unlike their social peers, had made the choice to study science by the age 12 and had remained committed to this decision. In addition, it is well recognized today, that there is a strong association between attitudes toward science and a child's performance in the science class. It was found, for instance, that children with more positive attitudes toward science showed increased attentiveness to classroom instruction and participated more in science activities (Germann, 1988). It is interesting to note that a stronger correlation between achievements in science and attitudes toward science was found for girls (Weinburgh, 1995). In the following, I will briefly describe some research which examined the influence of scientific field-trips on students' attitudes toward science.

On reviewing six studies conducted in informal settings, Falk (1983) found that they generally resulted in enjoyable and long-lasting memories. Harvey (1951, in Hofstein and Rosenfeld, 1996) found that an experimental group that underwent a series of geological field trips, out-performed the control group which discussed ecological concepts in a regular classroom, on the standard Caldwil and Curtis Scientific Attitude Test. This effect was attained even after short field visits. Jarvis and Pell (2002) examined attitude changes of children 10–11 years of age, after visiting the Challenger space simulation. They found that immediately after the Challenger experience, most of the children's attitudes were more positive. Twenty four percent of boys and girls became more positive about wanting to follow a scientific career in the future. The authors also found that this change in attitude was maintained to a certain extent for several months. These children also showed a statistically significant increase in science enthusiasm and an appreciation of its social context. In a later study Jarvis and Pell (2005) found that a visit to the UK National Space Center was an important factor in promoting higher interest in space for most children, and improved the children's attitudes toward science for some. One important result of this study was that the teachers' personal interest, preparation, actions during the visit, and follow-up were important factors in influencing children's short-and long-term attitudes. They also argue that the challenge of educators is to decrease the proportion of children, particularly girls, for whom the visit has little effect. They provide some suggestions to help teachers better exploit the potential of the scientific fieldtrip to impact positively on students' attitudes toward science. I would personally consider the out-of-school learning to be a success, even if non-formal learning environments only succeed in this domain, i.e., they only improve children's attitudes toward science and inculcate them with the passion to know more about science.

The Cognitive Aspect

While some researchers found that learning in scientific fieldtrips is ineffective (Anderson, 1994; Kubota and Olstad, 1991), others have argued that students constituted extremely valuable learning outcomes (Ayres and Melears, 1998; Ramey-Gassert *et al.*, 1994); outcomes that persist over time (Rennie, 1994; Wolins *et al.*, 1992). For instance, two studies conducted in the Singapore Science Center, one by Lam-Kan (1985) and the other by Fishon and Enochs (1987), found that students who interacted with the exhibit at the center, predominantly outperformed students who had no experience with the exhibition, regarding the concepts that underlined the exhibits. Realizing that children gain knowledge as a result of their visit to a science center is important. However, I feel that it is also important to seek out a theoretical explanation as to the potential for increased understanding of scientific concepts as a result of scientific fieldtrips. After all, no educator will stop taking his or her students on scientific fieldtrips just because a specific researcher found that students didn't gain the knowledge that they were expected to gain. Theoretical understanding might help educators to improve their designing of scientific fieldtrips to be more efficient. The Constructivism theory supports this.

Constructivism and Non-Formal Learning. Some researchers have argued that the hands-on activities in science centers, which are related to real-world objects and events, may be considered ideal learning environments according to the constructivism theory of learning (Falk *et al.*, 1986; Ramey-Gassert *et al.*, 1994). It is important to note that the views concerning learning and instruction, can sensibly be categorized in terms of *cognitive, social or cultural* constructivism (Windschitl, 2002). Cognitive constructivism is a system of explanations which deals with the manner in which learners, as individuals, adapt and refine knowledge (Piaget, 1971). I claim that meaningful learning is rooted in the idea that a person idiosyncratically restructures knowledge, actively basing it on his or her prior knowledge. As opposed to cognitive constructivism, social constructivism views knowledge as a primarily cultural product (Vygotsky, 1978). This is well expressed in the following citation:

> An interpersonal process is transformed into an intrapersonal one. Every function in the child's cultural development appears twice: first, on the social level, and later, on the individual level; first, between people (interpsychological) and then inside the child (intrapsychological) . . . . All the higher functions originate as actual relations between human individuals. (Vygotsky, 1978, p. 57)

Social constructivism, in the case of science museums, might be a good framework to help to understand what kind of learning processes occur during the dialogue among museum visitors and their manipulations with exhibits. Indeed, according to Gilbert and Priest (1997) "a group of visitors composed of individuals of varying experience of the phenomena involved, are able to share prior and present understanding through focused conversation, thus engaging in the social construction of knowledge" (pp. 750–751). This, according to the authors, is what makes museums so valuable in this regard. The authors argue that social context shapes individuals' mental models development. There is evidence in the literature that learning was indeed achieved through social interactions. For instance, Rahm (2004), based on the literature, argues that "through interaction of multiple voices (students and teachers) reflecting diverse interpretations, understandings, and personal experiences, knowledge is taken as essentially 'talked into being' " (p. 225). Moreover, Tunnicliffe (1997, 2000) who examined children's talking in museums, zoos, and botanical gardens, as well as Guberman and Van Dusen (2001), who examined children's investigations in a science discovery center, found that children, even without adult guidance, spontaneously engage in scientific thinking. However, it should be noticed that parents offer richer scientific learning opportunities to their children than their peers (Crowley and Callanan, 1998). One main component of social constructivism is the discourse that takes place among children, teachers, parents, or science center explainers. Gilbert and Priest (1997) went further, by identifying critical incidents, to analyze a discourse that took place during and after a visit by a class of 8–9-year olds at the Science Museum, London. The authors define a critical event to be:

> an event that is sufficiently coherent and apparently significant, as reflected in the discourse which takes place, to permit inferences to be made about the formation, use or development of mental models, as presented in the form of expressed models, by individuals in a social group. (p. 752)

Here are some types of critical incidents identified by the authors:

Discourse initiation

- Recognition of an object as being familiar
- Introduction of an element of surprise and providing an associated task
- Insertion of a question to focus pupils' attention

Discourse continuation

- Suggestion of ideas for post-visit activities
- Linking of generalized and particular
- Linking of objects
- Sustained attention provoked
- Successful consultation of text

Discourse closer

- Unsatisfactory nature of accompanying text.

By being aware of these, educators will be better able to plan conditions to foster efficient critical incidents that promote conceptual gain.

So far I referred to two aspects which, in my opinion, are crucial aspects of non-formal learning: the cognitive and affective. The literature offers some models for explaining out-of-school learning. I will now refer to two such models. One is the *contextual model* (Falk and Dierking, 2000); the other is the *three factors model* (Orion and Hofstein, 1994). I will first describe these models, and then critique them while arguing that a deep explanation should use the cognitive/affective division explicitly.

MODELS EXPLAINING SCIENTIFIC FIELDTRIP LEARNING

The Contextual Model

Learning is viewed by Falk and Dierking's (2000) *contextual model* as an effort to create meaning to survive and prosper within the world; an effort that is best viewed as a continuous, never-ending dialogue between the individual and his or her physical and socio-cultural environment. The authors identified eight key factors that affect learning within three contextual domains: personal, socio-cultural, and physical. They contended that if any of the eight principles are neglected, meaning making in the museum becomes more difficult.

The Personal Context

The personal context represents the sum total of personal and genetic history that an individual carries with him/her into a learning situation. From the personal context perspective one should expect learning to be influenced by:

1. Motivation and expectations
2. Prior knowledge, interests, and beliefs
3. Choice and control

The Social Context

The underlying assumption of the social context is that humans are extremely social in culture and hence, one should expect museums (and other forms of informal learning) always to be socio-culturally situated. Learning, according to this context, is influenced by:

4. Within-group socio-cultural mediation
5. Facilitated mediation by others

The Physical Context

The assumption here is that learning, which occurs within the physical environment, is in fact, always a dialogue with the environment. Thus, learning is influenced by the following environment components:

6. Advanced organizers and orientation
7. Design
8. Reinforcing events and experiences outside the museum

Orion and Hofstein's (1994) Three Factors Model

Orion and Hofstein's (1994) three factors model suggests that the following factors influence learning during scientific fieldtrips in natural environments:

- teaching factors, such as the location of the field trip in the curriculum structure, didactic methods, teaching and learning aids, and quality of teachers;
- field trip factors, such as the learning conditions for each learning station, duration and attractiveness of the trail, and weather conditions during the field trip; and
- student factors, such as previous knowledge of associated topics; previous acquaintance with area in question, previous experience with field trips, previous attitudes to subject matter, previous attitudes to field trips, and class characteristics (e.g. grade, size, and study major).

Critique of the Two Models

The teaching factor from Orion and Hofstein's model is not explicitly mentioned as one of the contexts in the contextual model. I believe that a model which can help teachers to better plan scientific fieldtrips should refer directly to the teaching context before, during, and after the visit. After all, it is too simplistic to see the fieldtrip as only occurring at the science center itself. It begins, in my opinion, with the preparation for the trip. Indeed, Folk and Dierking (2000) relate, in the physical context factor, to reinforcing events and experiences outside the museum; but this does not, in my opinion, explicitly put the teaching context where it is supposed to be. A good case to demonstrate this point can be taken from Lucas's (2000) paper. In this paper the author describes how he explained to the teacher that he wanted to attend the last lesson before the visit to the science center, to observe how she prepared the students for the visit. Her reply was, "you're probably already too late." At first he was surprised, but then he began to understand. The "visit" began several months before the actual trip. It also didn't end when the students left the science center. A good visit also includes activities that take place after the visit. A good model can not ignore the before and after visit activities. Orion and Hofstein, on the other hand, did not explicitly mention the social aspect as a factor which influenced the scientific fieldtrip, which is, in my opinion, rather surprising. Furthermore, it is my view that an efficient model is one which divides factors impacting the scientific fieldtrip into the cognitive and affective domains. These should not appear implicitly "in-between" the lines, but rather as categories according to which the other model's keys would fit. I present such an explanation here. Figure 2 illustrates my explanation. As described in Fig. 2, there are four factors which influence non-formal learning: personal, physical, social, and instructional. Each of these factors contains cognitive as well as affective components. For instance, the personal factor includes the child's prior knowledge, and belongs to the cognitive category. The personal factor also includes the child's agenda for the visit, his or her attitude toward science, as well as their efficacy beliefs. All of these factors belong to the affective domain. The social factor includes the interpersonal interaction which results in cognitive gain, as was explained previously by the cultural constructivism theory. Further, the social factor contains the influence of others (e.g. peers, teachers, family members, museum explainers) in the affective domains. Sometimes the interaction increases the motivation of the person to interact with a specific exhibition, which he or she might have otherwise ignored, had he/she been alone and vice versa. At first glance the reader might be surprised to find that the physical factor may also influence both the cognitive and affective domains; but, if he/she gives it some thought, he or she can realize that, for instance, the appearance of the exhibit, its color, the ease of manipulating it etc. may bear some influence on the affective domain. However, the degree by which one can manipulate the exhibit and how well it demonstrates scientific ideas, belong to the cognitive domain. Of course, the instructional factor also influences both on the cognitive and affective axes. As was previously discussed, the manner in which the teacher prepares students for the fieldtrip may

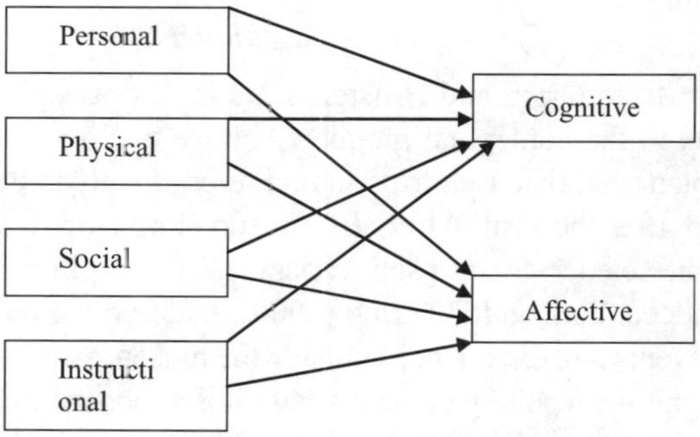

Figure 2. Factors influencing out-of-school learning each containing cognitive and affective domains

help children gain better understanding from the visit as well as prepare them emotionally.

I do not see my explanation as a kind of a model, but rather as something that organizes the factors already found in the literature, regarding the cognitive and affective domains. I would now like to discuss the *novelty phenomenon* from the standpoint of the previous explanation. It is an important phenomenon which is strongly associated with non-formal learning. Dealing with such a phenomenon through the eyes of this explanation might demonstrate its power.

The Novelty Phenomenon

Research on informal learning reveals a strong association between the novelty of an unfamiliar location stimuli, and visitor behavior (Falk, 1983; Falk and Balling, 1982), particularly in school groups. Balling suggests that: "The novel field situations produce an adaptation or adjustment on the part of the student which direct their behavior toward the environment and away from the structured learning activities" (p.128). According to Lucas (2000), high levels of novelty are reported with high levels of "off-task" behavior, at least in terms of teachers' objectives for students during a visit to a science museum or similar location.

Reviewing the literature, Burnett, Lucas, and Dooley (1996) identified three novelty-reduction approaches:

1. increasing students' familiarity with the physical location.
   In this regard, Orion (1993) argues that,

> Students should be prepared for the field trip. The more familiar they are with their assignment (cognitive preparation), with the area of the field trip (geographical preparation) and the kind of event in which they will participate (psychological preparation), the more productive the field trip will be for them (p. 326).

2. Insuring that students have the appropriate level of knowledge of the topics or focus of the exhibits/activities.
3. Providing preceding opportunities for students to practice relevant skills.

A unique manner of implementing these suggestions was reported in an interesting paper: *One Teacher's Agenda for a Class Visit to an Interactive Science Center* (Lucas, 2000). The author reports on how one teacher: Ms. Norton (whose views were also mentioned in the teachers' perspective section), invested considerable time and effort in the weeks leading up to the visit, to preparing the students for the visit by having them construct their own "exhibits." In her own words,

> The science centers got the equipment, and everything's set up, and lots of great learning experiences, but we're able to generate that ourselves probably on a lesser scale, so I just wanted to link the two, so that they understood that we had a mini-science-center. By the time we'd been through the process of building it, explaining it, showing how things work, and the way they do, that they would understand that's what the exhibits were for, it's not an entertainment. As much as it's fun, it's not like a time zone where they just go to get entertained. (Meg, 81097–234, in Lucas, 2000, p. 531)

This literature review reveals that the novelty effect influences children's performance on both the cognitive (this is also supported by the work of Kubota and Olstad, 1991; Riley and Kahle, 1995) as well as the affective (this is also supported in the work of Rennie; Riley and Kahle, 1995) learning outcomes. The students, when entering an unfamiliar location, might develop anxiety, and as a result be involved in off-task activities which may distance them from executing the learning tasks at hand. In addition, as was described, the teaching factor is strongly connected to the novelty factor. If teachers prepare their students for the fieldtrip beforehand, the children will feel much more comfortable, while experiencing less anxiety from their exposure to the new situation, and as a result will be more willing to learn. From what has just been said, to comprehend the novelty phenomenon, one should look through a pair of glasses which are comprised of one cognitive lens and one affective lens. In addition, the instructional factor should definitely be considered and seen as one that may help deal with the novelty phenomenon.

The benefits of non-formal learning both on the cognitive and affective axes also explain, using Howard Gardner's (Gardner, 1983, 1993) idea of multiple intelligences, why it may fit the needs of different people (Rennie and McClafferty, 1996).

The Multiple Intelligence Idea and Museum Learning

Howard Gardner's idea of multiple intelligences suggests a pluralistic view of the mind, with seven intelligences, rather than the traditional single intelligence implied by a single IQ score. Here is a brief description of the seven intelligences:

Interpersonal intelligence is concerned with the capacity to understand the intentions, motivations and desires of other people. It allows people to work effectively with others. Scientific fieldtrips usually require some degree of collaboration with others. Children usually work in groups to manipulate a specific exhibit. Thus such learning may fit those who have a strong interpersonal intelligence. Of course, it might develop such intelligence in those who weren't originally graced with it. In this case I would argue that because of the affective benefits of non-formal learning, the interpersonal intelligence might be addressed and undergo development.

Bodily-kinesthetic intelligence entails the potential of using one's whole body or parts of the body to solve problems. It is the ability to use mental abilities to coordinate bodily movements. Howard Gardner sees mental and physical activity as related. Of course, manipulating exhibits requires one to coordinate his or her body movement to perform a specific task and thus in such fieldtrips, those who possess a high level of such intelligence, might find themselves succeeding in the tasks even better than those who are usually considered the good science students. This might, of course, contribute to the improvement of one's self image in science.

Spatial intelligence involves the potential to recognize and use the patterns of wide space and more confined areas. According to Rennie and McClafferty (1996), the visitors in museums are usually involved with some kind of spatial or kinesthetic experience, and often work better with more than one person. In addition, I argue that in fieldtrips, the children have to navigate in unknown surroundings as well as manipulate 3-dimensional exhibits. These tasks require spatial abilities; thus, such learning might develop spatial intelligence.

Logical-mathematical intelligence consists of the capacity to analyze problems logically, carry out mathematical operations, and investigate issues scientifically. Many tasks in the scientific fieldtrip may require one to deal with problems, to provide an explanation for the unexpected behavior of a system, etc. These activities will probably attract those who have a high level of logical-mathematical intelligence, and, will contribute to its development in those who choose to deal with these kinds of tasks.

Linguistic intelligence involves sensitivity to spoken and written language, the ability to learn languages, and the capacity to use language to accomplish certain goals. This intelligence includes the ability to use language effectively to express oneself rhetorically or poetically as well as a means to remember information. Although science centers do not deal with languages, they do encourage one to express him or her self when explaining a phenomenon demonstrated by the exhibits. Thus, in some sense those who possess a high level of such intelligence might find themselves involved in an "explaining" role.

Intrapersonal intelligence entails the capacity to understand oneself, to appreciate one's feelings, fears and motivations. A fieldtrip is always an irregular occurrence. Thus, it might involve emotions toward the different gained experiences. The teachers may ask the students to think of things like: how they felt in the field trip and why? What parts they enjoyed and what parts they didn't. What did they learn and how? This means that non-formal learning might provide an opportunity to develop the intrapersonal intelligence, as well as to give those who have a high level of such intelligence a chance to demonstrate their ability.

Musical intelligence involves skill in the performance, composition, and appreciation of musical patterns. It encompasses the capacity to recognize and compose musical pitches, tones, and rhythms.

To summarize this point, non-formal learning can appeal to a range of intelligences, promoting the likelihood of engagement by people with different strengths and preferences for learning.

ON THE NEED TO BRIDGE

Thus far I discussed primarily non-formal learning. Indeed, most research of out-of-school learning relates to non-formal learning. But non-formal learning is only one possible form of out-of-school learning. Informal learning is another. It is not surprising that not much research has been carried out regarding informal learning. Places where informal learning takes place are out of teachers' and researchers' territory. One area of informal learning which drew the attention of some researchers is home learning, especially the connection between home and school learning. In this section, I describe some of this research. It is important for this chapter, which aims at bridging in and out of school learning. In addition, I suggest implementing the well known idea presented in the following phrase: "If Muhamed cannot come to the mountain, bring the mountain to Muhamed." By this I mean that if we wish to extend the use of scientific fieldtrips to science museums, and if bringing the children to the museum is complicated, then, in addition to the visit to the museums, the museums can visit the children. I will also present the idea of the "scientific kindergarten" which may also strengthen the connections between in and out of school learning.

Bridging Home and K-2 Classes

Solomon (2003) argues that "no one would deny the influence of home and families on the education of our children" (p. 219). Being aware of the significant influence that the home might have on learning, some educators sought after ways to establish stronger relationships between home and school. The SHIP (Solomon, 1993, 1994, 2003) project is one such an attempt. It aims at providing schools with banks of examples of simple activities which teachers could select as appropriate for children ages 5–10 years, to take home and carry out with their parents. The equipment used in the project is composed of simple objects and materials found in any household. Solomon (2003) claims that to understand science in the home, everything used should come from the home. The findings indicated that most parents showed real enjoyment of at least some of the activities provided by the project. In addition, in at least half of the investigations, the child had enough confidence to make some original contribution to the investigation. According to the author,

> In this way, they made the investigation at least partially their own, which rarely happens at school. They spoke easily with their parents and were encouraged, joked with, scolded, or ignored in a manner that clearly seemed familiar to them. (p. 229)

The author closed his paper saying that "a far greater reward from these activities with parents in their homes was the possibility of implanting the enjoyment of science into the home culture, and through this into the child's self image and future" (p. 231).

In another interesting study, Hall and Schaverien (2001), described what happened as children carried out scientific and technological inquiries, first as they were developed in school and then as they were pursued by children and families at home. The chosen topic was a flashlight, and the children, at the beginning session at school, demonstrated what they already knew about flashlights, how they worked and their everyday uses, recording these in the form of drawings and stories. The children were encouraged to ask questions and to develop their understanding of how flashlights might work. Each day, kits containing equipment such as batteries, wires, bulbs and switches were available for children to take home. The paper provides ample examples of learning situations which occurred at home. For instance, an example taken from one of the parents is the following story:

> that evening a friend called in — he's an engineer — and the three of them spent ages together, connecting circuits and blowing light bulbs. (Hall and Schaverien, 2000, p. 465)

On that occasion, George's father went further to challenge his son to making light bulbs of varying brightness; a challenge that George and his uncle, two weeks later, "were very enthralled — trying to solve the challenge" (p. 465). Another story tells how one parent, David's father, expressed surprise at his son's capability, explaining,

"This is great — I have an aversion to the science side of life — I think it is actually fear — I might get it wrong . . . . [David] has a 14-year-old brother who gave him a bit of help . . . he was playing with a little flashlight that he has." The authors did not know what David's brother did to help his understanding, but two weeks later David persisted in pursuing his own ideas, successfully connecting his complex circuit.

These two studies show that involving parents in science activities at school might result in good collaborative work in science and technology. This might have an important positive influence on students' attitudes and self image concerning science. It is important to mention, though, that in order to succeed in such efforts some guidance should be provided to families both pedagogically and scientifically.

Bringing Science Centers to the Class — Mobile Museums

As mentioned earlier, taking children to scientific centers might be problematic. First, science centers are not located everywhere. For instance, I was invited to conduct a workshop on *Inquiry Events* in Onsekiz Mart University in Çanakkale city. I was surprised to discover that they do not have any science centers in the entire area. This is only one example and there are probably many more places all around the world which do not have science centers. I thus argue that in places that don't have science centers, one can bring the science center to the class. It is my opinion that bringing the museum to the class is also good even in places where there are science centers. In such places, however, the role of the traveling museums will not be to replace the visit to museums. On the contrary, they can be used as a tool to prepare for the fieldtrip. An example of such a traveling museum is the Science on the Table program which was developed by Technocat in Israel. First, the table is big enough to hold many interesting apparatuses designed specifically for K-2 children, but at the same time, it is compact and small so that it can easily be moved to different corners of the classroom, and also can be transferred from school to school. The table includes different drawers, each containing different apparatuses on a certain topic. Figure 3 shows the table and the front drawers. The table also has big drawers on its sides.

Figure 3. The science table from the science on the table program.

Figure 4. Children playing with the balance scale from the Science on The Table program.

Here are some examples of its content: (1) An inclining plane which the children can play with and alter its angle. There are different planes made of different materials with different friction coefficients, such as carpet-like materials, veneer, or plastic. The children release different objects onto the plane, some with wheels and some without, and take note of the differences in the distances which they advance in relation to the type of material and the angle of the plane. They can measure the distance using the holes on the table. (2) A balance scale (see Fig. 4) and a set of four separate weights: a whole weight, three quarters, half, and a quarter. The children can hang the weights on opposite sides of the scale, and can see, for instance, that they need to hang two identical weights at the same distance from the center of the scale in order to balance it, but the balance will be broken if the two weights are placed at different distances. Also, they see that the closer the weight is to the center, the heavier the weight needed to balance with a weight further from the center on the other side. For instance, 1 quarter on the full distance, is balanced by 1 half at a halfway distance etc. (3) A set of cogwheels (see Fig. 5) which can be connected to the holes on the table. The children can investigate different reactions of the wheels to different setups. There are other apparatuses present in the table, including lenses and mirrors, a jukebox that works on cogwheels etc.

Creating Suitable Scientific Centers for K-2 Children

This chapter has shown the benefits of non-formal learning. I didn't, however, ignore the difficulties of such learning. For instance, I mentioned that research has found that visits to science centers often focus on facts or stories rather than substantial ideas or concepts. In addition, the vocabulary used during the tours might be too difficult for children to grasp. Also, there is little sensitivity, if any, to students' prior

Figure 5. A child playing with the cogwheels from the Science on the Table program.

experiences, as well as their individual and cultural differences. These factors might have an even greater impact on K-2 children than on older children. A creative way to solving such a problem is to build a special science center for small children. In Israel, there are several such centers. We used to call them Scientific Kindergartens. Now, unfortunately, they changed the name to Enrichment Centers. I think that the name Scientific Kindergartens is good since it emphasizes that this is a place especially for kids where science is the learning focus.

Characteristics of the "Scientific Kindergartens"

First, each scientific kindergarten serves several regular kindergartens from the surrounding areas — usually between 10–20 kindergartens. Each of the surrounding kindergarten's children visits the scientific kindergarten about 4–5 times a year, where each visit is devoted to activities on a specific topic. Typical topics are: sound and voices, optics, air, stones and rocks, electricity, and agriculture. The scientific kindergartens are usually highly equipped with advanced technology such as microscopes, dark rooms, pipes, and other instruments which are needed for their activities. In some scientific kindergartens the yards are also equipped so that science activities can be done outside. The person in-charge of the scientific kindergarten is usually a kindergarten teacher who has specialized in science throughout the years in-service courses. There are, of course, advantages and disadvantages to this situation. A person who is a kindergarten teacher by profession definitely understands what the children's needs are, as well as the kindergarten teachers' needs. That person is also aware of the pedagogical methods appropriate for K-2 children and knows how to present the scientific topics to the children. Moreover, it is also expected that good working connections may develop between the scientific kindergarten teacher and the children's preschool teacher. The scientific kindergarten teacher meets the regular kindergarten teacher at the beginning of the year. In this meeting she describes what the topics, which are to be covered during the school year, are. She also provides some suggestions for preparing the kids for the meetings. In addition, after the children attend the scientific kindergarten, their teacher is provided with follow-up activities which she can conduct at her kindergarten. Furthermore, the scientific kindergarten teachers also visit the children in their kindergarten during the year. She may see a scientific activity during these visits or even conduct one herself. In such a manner she becomes a familiar figure in the lives of the children, as well as the center itself. Thus, the novelty effect is expected to minimize. On the other hand, despite participating in some scientific courses, the scientific kindergarten teachers still lack scientific knowledge. To help these teachers they can usually be assisted by a scientist who serves as a kind of consultant for the center.

DISCUSSION

The first chapter of this book provided some justifications as to the importance of beginning the constitution of the child's scientific foundations as early as kindergarten. The subsequent chapters focused on approaches which I found suitable for science teaching at this age. Although these approaches are not restricted to only formal learning, no specific attention was given to the out-of-school learning. Without obtaining a specific and explicit understanding of out-of-school learning, which the present chapter deals with, one can not, in my opinion, fully comprehend K-2 science teaching. I chose this chapter for the ending of this book not because I hold out-of-school learning to be less important than formal learning, but because I thought that being equipped with the potential of formal learning, as well as being acquainted with the different approaches on how to teach it, is a necessary background before one carries on to the less-formal nature of non-formal and informal learning.

In understanding the importance of dealing with out-of-school learning, one should refer to the unfortunate fact that schools alone have not usually been successful in creating scientifically literate school leavers. As was discussed in this chapter, as well as is stated by Jarvis and Pell (2002), "the process of enabling young children to start a lifelong interest and understanding of science in the wider world may be improved by the provision of out-of-school science experiences" (p. 980). The authors find support for this view in the following citation:

> Unless the young people of the twenty-first century appreciate the importance of science, we stand no chance whatsoever of economic, social or cultural survival. In my view, science museums and science centers must play an appropriately active part in the educational programme on which this survival depends. (H. Kroto, joint winner of the 1996 Nobel Prize for chemistry, 1997, p. 14)

Howard Gardner, in his book *The Unschooled Mind, How Children Think and How Schools Should Teach*, even further emphasizes how important the science museums' roles are by his envision of the following learning environment:

> Imagine an educational environment in which youngsters at the age of seven or eight, in addition to — or perhaps instead of — attending a formal school, have the opportunity to enroll in a children's museum, a science museum, or some kind of discovery center or Exploratorium. As part of this educational scene, adults are present who actually practice the disciplines or crafts presented by the various exhibitions. Computer programmers are working in the technology center, zookeepers and zoologists are tending the animals, workers from a bicycle factory assemble bicycles in front of the children's eyes, and a Japanese mother prepares a meal and carries out a tea ceremony in the Japanese house. Even the designers and the mounters of the exhibitions ply their trade directly in front of the observing students. (p. 200)

The author continues and asks "Would we not be consigning students to ruination if we enroll them in museums instead of schools?", and he answers, "I believe we would be doing precisely the opposite. Attendance in most schools today does risk ruining the children."

I do not agree that sending children to schools today risks ruining them. I have found teachers in many cases to be doing wonderful work and advancing children cognitively and emotionally. They do, of course, still have room for improvement. This, however, does not mean that parents are doing something wrong by sending their children to school. I also think that such sharp criticism contributes to decreasing the teachers' status, which already suffers tremendously. This entire book was written from the point of view that teachers are doing important and crutial work, and invest a lot of effort: we educational researchers should help them to tunnel their efforts more efficiently. Also, I do not think that museums should replace schools in spite of their advantages. After all, reviewing the literature, this chapter revealed that museums have much to improve themselves. Also, one important factor of fieldtrips over schools is the fact that it somewhat changes the routine. So, I herein call that we should not ruin breaking the routine. It is my view that non-formal institutions should not replace schools. Schools should remain schools; but educators must construct bridges so that out-of-school learning, be it informal or non-formal, is better connected to the in-school learning. I am sure that teachers do not fully understand the role of out-of-school learning in science teaching. This is not surprising. Even searching *The Hand Book of Research on Science Teaching and Learning (Gable, 1994)* I did not find, to my astonishment, any explicit treatment of out-of-school learning phenomena. Hence, in the eyes of some educators, these phenomena might be interpreted as unimportant. Being aware of its importance, as well as being equipped with some suggestions as to how to bridge in and out-of-school learning, I hope that this book will bring the latter to where it belongs, among others — in the school.

I suggested in this chapter that four factors influence learning in non-formal learning environments: personal, physical, social, and instructional. Each contains cognitive and affective components. I argued that in order to ensure an efficient scientific fieldtrip, one should appropriately treat each of the above factors, on both the cognitive and the affective levels. To bring theory into practice and to apply what has been discussed in the chapter, one should consider the following when designing and executing scientific fieldtrips:

Implication to Education

- Decide what the purpose of the scientific fieldtrip is. For instance, is it a kind of enrichment experience that is not connected to the curriculum? Is it connected? In this case the teacher should decide whether its role is to introduce or motivate a learning topic, to summarize it, or to deepen and extend it. The purpose, of course, directs the way in which it will be conducted.
- Visit the fieldtrip location beforehand. Talk with the people in-charge of the educational program to inform them about the purpose of the visit and your expectations. Ask them whether they have any suggestions for activities you can do in the class before the visit and afterwards which, of course, fit the fieldtrip purpose.
- Share the purpose of the visit with the children before the visit and share your expectations of them. You can also ask the children whether they have any expectations of their own. In this regard it must be clear to the children that the visit is a learning experience.
- To decrease the novelty phenomena, the children can be presented with the structure of the day. In addition, the children may enter to the location's Internet site and become acquainted, to some extent of course, with the environment before the fieldtrip. In such a way they may feel safe not being afraid about being lost or not knowing what to do. This will also decrease concerns from the teacher's side.
- Conduct the relevant scientific activities in the class before going to fieldtrip. This is important because in this way the children will acquire both the skills and background knowledge they need in order to better benefit from the new experiences.
- Always provide some tasks to be conducted in the fieldtrip. This is very important because such tasks may help the children to notice things that could otherwise be ignored. It is suggested that the tasks be open-ended and require observation, discussion, and deduction of ideas or principles rather than a focusing on recording of factual information. Also, it is important not to overwhelm the children with too many tasks. In this regard more is not always better. In addition, it is important to bear in mind that the child should also have the opportunity to have free choices both in what exhibits or activities he or she wants to participate in and in what manner they want to conduct it.
- Involve parents and encourage them to join the trip. Remember, it was found that adults' help might stimulate them and lead to longer and deeper involvement with the exhibits. For this purpose, of course, parents with some scientific background might be a good fit.
- It is suggested that schools prepare some activities in advance; activities that may be good for parents to join. In addition, there should be some guidance on how parents can continue those activities at home. Such activities might encourage parents to conduct scientific activities with their children. This might have a positive affect on the child's motivation, attitudes, and self image concerning science.
- Schools and museums should cooperate and bring some scientific activities into the classroom. In such a way there might be more and stronger interactions between schools and museums.

- I also call for broadening the *scientific kindergarten* model. Such centers, that might even be part of science museums, might better fit the activities to the children's needs.
- In-service courses for teachers are needed first to increase their awareness of what out-of-school learning environments may offer and second to teach them how to execute scientific fieldtrips more effectively.

Considering the advantages and disadvantages of out-of-school learning, I find it has great potential to help people to learn, appreciate, and develop positive feelings toward science. This means that such a kind of learning has an impact on both the cognitive and affective axes. However, as was shown in the chapter, this potential is not fully exploited. The suggestions I have made might contribute to improve this situation. I also think that the out-of-school learning should be treated more in the science education literature. First, hand-books should handle such learning. Second, most of the research was done regarding non-formal learning environments, especially science centers and museums. We do need to know more about children's learning processes at home, with their parents. We need to know more about what they learn in playgrounds, nature excursions etc. Learning and teaching should be seen in a holistic way; hence, we should immerse ourselves, and I hope this book contributes in this matter, in understanding of in-school learning, as well as out-of-school learning.

MATOME

In the Introduction I mentioned the Japanese term, *tsunami*. To close the circle I chose to give the end-piece of my book another Japanese name, *matome*. *Matome*, for the Japanese, is the part of a lesson in which the main points of the lesson are 'summed up' or 'pulled together', or in other words, the summary.

In the first chapter, it was argued that one of the six justifications to science education as early as childhood is that children are capable to understand complex concepts and are even able, to some extent, to connect theory and evidence, i.e. think scientifically. Hence, it was argued that educators ought to expose children to situations in which those abilities find fertile ground to grow. *Why* educators fail to design such scientific activities was not really discussed. This was partly because we had not yet developed the broad theoretical background regarding K-2 science education, which was one of the goals of this book. Now, however, let me consider this issue briefly, not only by itself, but also as a first step toward 'pulling together' what we have done in this book. In her article "Reassessment of Developmental Constraints on Children's Science Instruction," Metz (1995) argues that as a result of the following wrong assumptions, the elementary science curricula accepted in most schools today is far, far below the cognitive abilities of children:

1. Logical mathematical structures of seriation and classification constitute core intellectual strengths of concrete operational elementary school children. These enable them to organize concrete objects using seriation and classification. Therefore, observation, ordering, categorization, and corresponding inferences and communications, are appropriate objectives that should be emphasized in science instruction at the elementary school level.
2. Elementary school children are concrete operational who are "concrete thinkers," whose reasoning is tied to concrete objects and their manipulation. Abstractions, ideas not tied to concrete situations, are beyond their grasp. Therefore we need to restrict children's science curricula to concrete and "hands-on" activities and post-pone abstractions until higher grade levels.
3. The logic of experimental control and inference does not emerge until adolescence. This stems from the belief that formal operational thought, which in contrast to concrete operational thought is more systematic, less egocentric, and more abstract, is developed only in adolescence. Therefore scientific investigations in the form of planning and implementing experiments and drawing inferences from the complex of outcomes should be largely postponed.

According to Metz (1995),

> These developmental constraints on children's science instruction are not supported by the Piagetian or non-Piagetian developmental literature. Few cognitive psychologists believe that seriation and classification constitute core intellectual strengths of elementary school children. These children are capable of grasping at least some abstract ideas. They can engage in scientific inquiry and infer new knowledge on the basis of their experimentation. Thus, it is not necessary to emphasize the process of observing, ordering, and categorizing the directly perceivable and concrete, while relegating scientific investigation to later years. This developmental literature indicates that elementary school children are actually capable of a much richer scientific inquiry than these assumptions imply (p.120).

In a later article, *Scientific Inquiry Within Reach of Young Children*, Metz (1998) argues that current instructional interventions demonstrate the possibility of strengthening children's scientific inquiry through the support of suitable instruction within different aspects of the scientific process.

From what has been said, it seems that there is a serious gap between what children are capable of doing and understanding, and the experiences they get in school, not to mention, of course, in kindergarten. This book is addressed specifically to this problem. It first discusses the importance of science education for children. Then it provides theoretical explanations as to how one can teach science in a manner fitting children's cognitive abilities, hence possessing greater potential to contribute to their cognitive development. It was suggested, moreover, that especially in the case of science, preschool teacher's needs should also be considered.

I recently participated in the 11th biennial EARLI conference in Nicosia, Cyprus which was held on August 23–27, 2005. In her keynote address, entitled *Does Learning Develop?*, the distinguished researcher, Deana Kuhn, argued that older children and adolescents have naturally more experience and definitely more time and opportunities to learn than younger children. As a result, she argued that they certainly know more. This is, of course, one difference between the two groups. However, her question was: Does the learning process *itself* differ with age? Her answer to this question was that conceptual learning, one which involves change in understanding, requires cognitive engagement on the part of the learner, and hence an executive that must allocate, monitor, and otherwise manage the mental resources that are involved. These executive functions, and the learning that requires them, do show evidence of development. In addition, meta-cognitive operators become more prominent with age. Thinking of my book, after her lecture, I approached her and asked whether she thinks that by appropriate scaffolding we can enable children to develop those executive control functions, as well as the meta-cognitive operators. Her response was a resounding YES. Kuhn's lecture, then, trenchantly reinforced the thesis which I have maintained throughout this book, namely, that good science education can and should start early in life.

This book has proposed some justification to K-2 science education and has offered some approaches and methods to teaching it, but we are still just at the beginning of the road. Many questions have yet to be addressed. For instance, why do some science activities work better than others with children? How can we prepare teachers for science education in kindergarten? How widely and in what way do teachers pass on their scientific knowledge and skills to their children? What kind of activities might advance executive control functions in children? What difficulties do children have in understanding scientific knowledge and acquire scientific skills. What activities are required to develop meta-cognitive operators in children? How can we analyze whether scientific activities efficiently scaffold scientific knowledge and scientific reasoning? In addition, how might educators best invest effort to build science curricula that take into account the points discussed in this book?

The questions just mentioned are questions for the future. But thinking about the future should not mean forgetting the past. So I would like to close this book with a quotation from John Dewey, who like me, dedicated so much of his efforts to both scientific thinking and to children. Thus, almost one hundred years ago, Dewey wrote:

> . . . the native and unspoiled attitude of childhood, marked by ardent curiosity, fertile imagination, and love of experimental inquiry, is near, very near, to the attitude of the scientific mind. (Dewey, 1910, p. iii)

BIBLIOGRAPHY

American Association for the Advancement of Science (1993). *Benchmarks for Science Literacy*. New York: Oxford University Press.

Anderson, D. (1994). *The effect of pre-orienting year eight students to the informal learning environment of a science museum on cognitive learning*. Unpublished master's thesis, Queensland University of Technology.

Ansbacher, T. (1998). John Dewey's experience and education: Lessons for museums. *Curator*, 41: 36–49.

Appleton, K. (2002). Science activities that work: Perceptions of primary school teachers. *Research in Science Education*, 32: 393–410.

Aristotle (1941). *The Basic Works of Aristotle*, Richard McKeon (Ed.). New York: Random House, Inc.

Ausubel, D. P. (1968). *Educational Psychology: A Cognitive View*. New York: Holt, Rinehart & Winston.

Ayres, R. and Melear, C. T. (1998). *Increased learning of physical science concepts via multimedia exhibit compared to hands-on exhibit in a science museum*. Paper presented at the annual meeting of the National Association for Research in Science Teaching (NARST), San Diego, CA.

Barlex, D. and Pitt, J. (2002) The relationship between Design and Technology and Science. In Owen-Jackson, G. (Ed.), *Teaching Design and Technology in Secondary Schools — A Reader*. London: Routledge Falmer and The Open University, pp 177-192.

Bauer, H. H. (1994). *Scientific Literacy and the Myth of the Scientific Method*. Urbana: University of Illinois Press.

Begley, S. (1996). Your child's brain. *Newsweek*, February 19: 41–46.

Ben-Zvi Assaraf, O. and Orion, N. (2005). Development of system thinking skills in the context of earth system education. *Journal of Research in Science Teaching*, 42: 518–560.

Black, P. and Harlen, W. (1993). How can we specify concepts for primary science? In Black, P. J. and Lucus, A. M. (Eds), *Children's Informal Ideas in Science*. London: Routledge, pp. 208–229.

Black, P., & Harrison, G. (1992). In McCormick, R et al (Eds). *Teaching and learning technology*. Wokingham, England: Addison-Wesley Publishing company, in association with the Open University, Milton Keynes.

Bloom, B. S. (1954). *Taxonomy of Educational Objectives. Handbook 1: Cognitive Domain*. New York: Logmans, Green & Co.

Bloom, B. (1956). *Taxonomy of Educational Objectives: Handbook-I. Cognitive Domain*. New York: David Mckay.

Boden, M. A. (1999). Computer models of creativity. In Sternberg, R. J. (Ed.), *Handbook of Creativity*. Cambridge: Cambridge University Press.

Bodner, G. (1986). Constructivism: A theory of knowledge. *Journal of Chemical Education*, 63: 873–878.

Borun, M. and Dritsas, J. (1997). Developing family-friendly exhibits. *Curator*, 40: 178–196.

Bourdieu, P. (1992). *Language and Symbolic Power*. Cambridge, MA: Harvard University Press.

Bowden, J., Dall'Alba, G., Martin, E., Laurillard, D., Marton, F., Master, G., Ramsadan, P., Stephanous, A., and Walsh, E. (1992). Displacement, velocity, and frames of reference: Phenomenongraphic studies of students' understanding and some implications for teaching and assessment. *American Journal of Physics*, 60: 262–269.

Boyle, D. G. (1971). *Language and Thinking in Human Development*. London: Hutchinson University Library.

Brooks, L. R. (1968). Spatial and verbal components of the act of recall. *Canadian Journal of Psychology*, 22: 349–368.

Brown, A. L. (1990). Domain-specific principles affect learning and transfer in children. *Cognitive Science*, 14: 107–133.

Brown, A. L. and Campione, J. C. (1994). Guided discovery in a community of learners. In K. McGilly (Ed.), *Classroom Lessons: Integrating Cognitive Theory and Classroom Practice*. Cambridge, MA: MIT Press/Bradford Press, pp. 229–270.

Brown, S. I. and Walter, M. I. (Ed.) (1990). *The Art of Problem Posing*. 2nd ed. Hillsdale, NJ: Lawrence Erlbaum Associates, Publishers.

Brown, J. S., Collins, A., and Duguid, P. (1989). Situated cognition and the culture of learning. *Educational Researcher*, 18(1): 32–42.

Bruce, B. C., Bruce, S., Conrad, R., and Huang. H. (1997). Collaboration in science education: University science students in the elementary school classroom. *Journal of Research in Science Teaching,* 34(1): 69–88.

Bruner, J. S. (1960). *The Process of Education*. Cambridge, MA.: Harvard University Press.

Bruner, J. (1963). *The Process of Education*. New York: Vintage.

Burnett, J. R., Lucas, K. B., and Doley, J. H. (1996). Small group behavior in a novel field environment: Senior science students visit a marine theme park. *Australian Science Teachers' Journal*, 42: 59–64.

Butt R., Raymond D., and Yamaguishi L. (1982). Autobiographic praxis: Studying the formation of teachers' knowledge. *Journal of Curriculum Theorizing*, 7: 87–164.

Cajas, F. (1999). Public understanding of science: Using technology to enhance school science in everyday life. *International Journal of Science Education*, 21: 765–773.

Carey, S. (1978). The child as word learner. In Halle, M., Bresnan, J., and Miller, G. (Eds), *Linguistic Theory and Psychological Reality*. Cambridge, MA, pp. 264–293.

Carroll E. (1988). The role of tacit knowledge in problem solving in the clinical setting. *Nurse Education Today*, 8: 140–147.

Carson, R. (1984). *The Sense of Wonder*. New York: Harper & Row Publishers.

Cazden C. B. (1979). Peekaboo as an instructional model: Discourse development at home and at school. In *Papers and Reports On Child Language Development* (No. 17). Palo Alto, CA: Stanford University, Department of Linguistics.

Champagne, D. W. (1975). The Ontario science center in Toronto: Some impressions and some questions. *Educational Technology*, 15(8): 36–39.

Champagne, A. B. and Klopfer, L. E. (1977). A sixty-year perspective on three issues in science education. *Science Education*, 61: 431–452.

Chan, C., Burtis, J., and Bereiter, C. (1997). Knowledge-building as a mediator of conflict in conceptual change. *Cognition and Instruction*, 15: 1–40.

Cho, H., Kahle, J. B., and Nordland, F. H. (1985). An investigation of high school biology textbooks as a source of misconceptions and difficulties in genetics and some suggestions for teaching genetics. *Science Education*, 69: 707–719.

Clark, E. V. (1983). Meanings and concepts. In Mussen, P. H. (Ed.), *Handbook of Child Psychology.* New York: John Wiley, pp. 787–840.

Clement, J. (1982). Students' preconceptions in introductory mechanics. *American Journal of Physics*, 50: 66–71.

Clement, J. (1987). Overcoming students' misconceptions in physics: The role of anchoring intuitions and analogical validity. In Novak, J. D. (Ed.), *Proceedings of the Second International Seminar Misconceptions and Educational Strategies in Science and Mathematics,* 3 (Vol. 1). Ithaca, NY: Cornell University, pp. 84–97.

Clement, J. (1988) *Imagery in problem solving in physics*. Paper presented at the Workshop on Cognition in Science Learning, Tel-Aviv, Israel.

Clement, J. (1994). Use of physical intuition and imagistic simulation in expert problem solving. In Tirosh, D. (Ed.), *Implicit and Explicit Knowledge*, Norwood, NJ: Ablex Publishing, pp. 204–244.

Cobern, W. W. and Loving, C. C. (2002). Investigation of preservice elementary teachers? thinking about science. *Journal of Research in Science Teaching*, 39(10): 1016–1031.

Cohen G. (1996). *Memory in the Real World*. 2nd ed. Hove, East Sussex, UK: Psychology Press.

Cohen, R., Eylon, B., and Ganiel, U. (1983). Potential difference and current in simple electric circuits: A study of students' concepts. *American Journal of Physics*, 51: 407–412.

Collins, A. and Quillian M. (1969). Semantic hierarchies and cognitive economy. *Journal of Verbal Learning and Verbal Behavior*, 8 (2): 7–240.

Court, A. W. (1998). Improving creativity in engineering design. *European Journal of Designing Education*, 23: 141–154.

Cox-Petersen, A. M., Marsh, D. D., Kisiel, J., and Melber, L. M. (2003). Investigation of guided school tours, student learning, and science reform recommendations at a museum of natural history. *Journal of Research in Science Teaching*, 40(2): 200–218.

Craver-Lemley, C., and Reeves, A. (1992). How visual imagery interferes with vision. *Psychological Review*, 99: 633-649.

Crawford, B. A., Krajcik, J. S., and Marx, R. W. (1999). Elements of a community of learners in a middle school science classroom. *Science Education*, 83: 701–723.

Crawley, F. E. and Black, C. (1992). Causal modeling of secondary students' intention to enroll in physics. *Journal of Research in Science Teaching*, 29: 585–599.

Cripps, R.J. and Smith, F. (1993). An information system to support the design activity. *Journal of Engineering Design*, 4: 3–10.

Crismond, D. (2001). Learning and using science ideas when doing investigate-and-redesign tasks: A study of naïve, novice, and expert designers doing constrained and scaffolded design work. *Journal of Research in Science Teaching*, 38: 791–820.

Crowley, K. and Callanan, M. A. (1998). Identifying and supporting shared scientific reasoning in parent-child interactions. *Journal of Museum Education*, 23: 12–17.

Csikszentmihalya, M. and Hermanson, K. (1995). Intrinsic motivation in museums: What makes visitors want to learn? *Museum News*, 74: 34–37, 59–61.

Cunningham, C. F. and Blankenship, J. W. (1979). Preservice elementary science teachers' self concerns. *Journal of Research in Science Teaching*, 16: 419–425.

de Vries, M. J. (1994). Technology education in Western Europe. In Layton, D., (Ed), *Innovations in science and technology education* (Vol. 5). UNESCO Publishing.

de Vries, M. J. (2005). *Teaching About Technology — An Introduction to the Philosophy of Technology for Non-philosophers*. Dordrecht, The Netherlands: Springer.

Department of Education and Science/Welsh Office. (1990). *National curriculum design and technology for ages 5-16*. London: Her Majesty's Stationery Office.

Dewey, J. (1910). *How We Think*. Lexington, Mass: D.C. Heath, (1910)

Dewey, J. (1910/1997). *How We Think*. Mineola, NY: Dover Publication, Inc.

Dewey, J. (1913). *Interest and Effort in Education*. Boston: Houghton Mifflin.

Dewey, J. (1916). *Democracy and Education*. New York: Macmillan, pp. 256–261.

Dewey, J. (1916/1966). *Democracy and Education — An Introduction To the Philosophy of Education*. New York: The Free Press.

Dierking, L. D. (1991). Learning theories and learning style: An overview. *Journal of Museum Education*, 16: 4–6.

Dierking, L. D. and Falk, J. H. (1994). Family behavior and learning in informal science settings: A review of the research. *Science Education*, 78(1): 57–72.

Dillon, J. T. (1990). *The Practice of Questioning*. London: Routledge.

Dori, Y. J. and Herscovitz, O. (1999). Question posing capability as an alternative evaluation method: Analysis of an environmental case study. *Journal of Research in Science Teaching*, 36: 411–430.

Drake, S. (1978). *Galileo at Work: His Scientific Biography*. Chicago: University of Chicago Press.

Driver, R. and Bell, B. (1986). Students' thinking and the learning of science: A constructivist view. *School Science Review*, 67: 443–456.

Driver R., Guesne, E., and Tiberghien, A. (1985). Some features of children's ideas. In Driver R., Guesne, E., and Tiberghien, A. (Eds), *Children's Ideas in Science*. Philadelphia: Open University Press, pp. 193–201.

Drucker, P. F. (1961). The technological revolution: Notes on the relationship of technology, science and culture. *Technology and Culture*, 2 (4): 342–351.

Druyan, S. (1997). Effect of the kinesthetic conflict on promoting scientific reasoning. *Journal of Research in Science Teaching*, 34: 1083–1099.

Dunbar, K. and Klahr, D. (1989). Developmental differences in scientific discovery strategies. In Klahr, D. and Kotovsky, K. (Eds), *Complex Information Processing: The Impact of Herbert A. Simon*. Hillsdale, NJ: Erlbaum, pp. 109–143.

Duran, D. and Monereo, C. (2005). Styles and sequences of cooperative interaction in fixed and reciprocal peer tutoring. *Learning and Instruction*, 15: 179–199.

Edelson, D. C. (2002). Design research: What we learn when we engage in design. *Journal of the Learning Sciences*, 11, 105-121.

Einstein, A. and Infeld, L. (1938). *The Evolution of Physics*. New York: Simon & Schuster.

Eshach, H. (2003). Small-group interview-based discussions about diffused shadow. *Journal of Science Education and Technology*, 12: 261–275.

Eshach, H. and Bitterman, H. (2002). From case-based reasoning to problem-based learning. *Academic Medicine,* 78: 491–496.

Eshach, H. and Schwartz, J. (2004). *Middle school students' preconceptions of sound*. Paper presented at the annual meeting of the National Association for Research in Science Teaching (NARST), Vancouver, BC, Canada, April 1–3, 2004.

Eysenck, M. W. and Keane, T. M. (1995). *Cognitive Psychology: A Student's Handbook*. 3rd ed. Hove, East Sussex, UK: Psychology Press.

Falk, J. H. (1983). Field trips: A look at environmental effects on learning. *Journal of Biological Education*, 17: 137–141.

Falk, J. H. and Adelman, L. M. (2003). Investigating the impact of prior knowledge and interest on aquarium visitor learning. *Journal of Research in Science Teaching*, 40(2): 163–176.

Falk, J. H. (2004). The director's cut: Toward an improved understanding of learning from museums. *Science Education*, 88(11): S83–S96.

Falk, J. H. and Balling, J. D. (1982). The field trip milieu: Learning and behavior as a function of contextual events. *Journal of Educational Research*, 76: 22–28.

Falk, J. H. and Dierking, L. D. (2000). *Learning From Museums: Visitors Experiences and Their Making of Meaning*. Walnut Creek, CA: Altamira Press.

Falk, J. H. and Storksdieck, M. (2005). Using the contextual model of learning to understand visitor learning from a science center exhibition. *Science Education*, 89(5): 1–35.

Falk, J. H., Koran, J. J. Jr., and Dierking, L. D. (1986). The things of science: Assessing the learning potential of science museums. *Science Education*, 70: 503–508.

Feher, E. and Rice, K. (1988). Shadows and anti-images: Children's conceptions of light and vision. *Science Education*, 72: 637–649.

Feibleman, J. K. (1972). Pure science, applied science and technology: An attempt at definitions. In C. Mitcham and R. Mackey (Eds.), *Philosophy and technology*. New York: Free Press.

Fensham, P. J. and Gardner, P. L. (1994). Technology education and science education: A new relationship? In Layton, D. (Ed.), *Innovations in Science and Technology Education*, 5: 159–170.

Ferguson, E. L. and Hegarty, M. (1995). Learning with real machines or diagrams: Application of knowledge to real-world problems. *Cognition and Instruction*, 13(1): 129–160.

Feynman, R. (1988). *What do You Care What Other People Think?* New York: Bantan Books.

Feynman, R. P. (1989). *What do You Care What Other People Think? Further Adventures of a Curious Character*. New York: Bantam Books.

Finke, R. A. (1989). *Principles of Mental Imagery*. Cambridge, MA: MIT Press.

Fleer, M. (1992). Identifying teacher-child interaction which scaffolds scientific thinking in young children. *Science Education*, 76: 373–397.

Franz, J. R. and Enochs, L. G. (1982). Elementary school science: State certification requirments in science and their implications. *Science Education*, 66: 287–292.

Gable, (Ed.) (1994). *The Hand Book of Research on Science Teaching and Learning*. New York: Macmillan Publishing Company, National Teachers Science Association.

Galili, I. and Hazan, A. (2000). Learners' knowledge in optics: Interpretation, structure and analysis. *International Journal of Science Education*, 22: 57–88.

Gange, R. M. (1974). *The Conditions of Learning*. 2nd ed. New York: Holt, Rinerhart & Winston.
Gardner, H. (1983). *Frames of Mind*. New York: Basic Books.
Gardner, H. (1991). *The Unschooled Mind*. New York: Basic Books.
Gardner, H. (1993). *Multiple Intelligences: The Theory of Multiple Intelligences*. New York: Basic Books.
Gardner, H. (1999). *The Disciplined Mind — What All Students Should Understand*. New York: Simon & Schuster.
Gardner, A. L. and Cochran, K. F. (1993). Critical issues in reforming elementary teacher preparation in mathematics and science: *Conference proceedings*. Greeley, CO: Center for Research on Teaching and Learning.
Garrett, R. M., Satterly, D., Gil Perez, D., and Martinez Torregrosa, J. (1990). Turning exercises into problems: An experimental study with teachers in training. *International Journal of Science Education*, 12: 1–12.
Gelman, S. A. and Markman, E. M. (1886). Categories and induction in young children. *Cognition*, 23: 183–208.
Gerber, B. L., Marek, E. A., and Cavallo, A. M. L. (2001). Development of an informal learning opportunities assay. *International Journal of Science Education*, 23(6): 569–583.
Germann, P. J. (1988). Development of the attitude toward science in school assessment and its use to investigate the relationship between science achievement and attitude toward science in school. *Journal of Research in Science Teaching*, 25: 689–703.
Gibson, S. and Dembo, M. H. (1984). Teacher efficacy: A construct validation. *Journal of Educational Psychology*, 76: 569–582.
Gilbert, J. and Priest, M. (1997). Models and discourse: A primary school science class visit to a museum. *Science Education*, 81: 749–762.
Gilbert, J. K. and Reiner, M. (2000). Thought experiments in science education: Potential and current realization. *International Journal of Science Education*, 22: 265–283.
Glover, J. A., Ronning, R. R., and Reynolds, C. R. (Eds). (1989). *Handbook of Creativity*. New York: Plenum.
Gowan, J. C. (1978). Incubation, imagery, and creativity. *Journal of Mental Imagery*, 2(Spring): 23–31.
Griffin, J. (2004). Research on students and museums: Looking more closely at the students in school groups. *Science Education*, 88(11): S59–S70.
Griffin, J. and Symington, D. (1997). Moving from task-oriented to learning-oriented strategies on school excursions to museums. *Science Education*, 81: 763–779.
Guberman, S. R., and Van Dusen, A. (2001). Children's investigations in a science center. Paper presented at the American Educational Research Association, Seattle.
Guesne, E. (1985). Light. In Driver, R., Guesne, E., and Tiberghien, A. (Eds), *Children's Ideas in Science*, Open University Press, Milton Keynes, U.K./Philadelphia, pp. 10–32.
Gustafson, B. J. and Rowell, P. M. (1995). Elementary preservice teachers: Constructing conceptions about learning science, teaching science and the nature of science. *International Journal of Science Education*, 17: 589–605.
Hadamard, J. (1945). *The Psychology of Invention in the Mathematical Field*. Princeton, NJ: Dover publishers.
Hall, R. and Schaverien, L. (2001). Families' participation in young children's science and technology learning. *Science Education,* 85 (4): 454–481.
Haigh, M., France, B., and Forret, M. (2005). Is "doing science" in New Zealand classrooms an expression of scientific inquiry? *International Journal of Science Education*, 27: 215–226.
Halloun, I. A. and Hestenes, D. (1985). Common sense concepts about motion. *American Journal of Physics*, 53: 1056–1065.
Hand, B., Wallace, C. S., and Yang, E.M. (2004). Using the science writing heuristic to enhance learning outcomes from laboratory activities in seventh grade science: Quantitative and qualitative aspects. *International Journal of Science Education*, 26: 131–149.
Harlen, W. (1997). Primary teachers' understanding in science. *Research in Science Education*, 27: 323–337.
Harvey, H. W. (1951). An experimental study of the effects of field trips upon the development of scientific attitudes in a 9th grade general science class. *Science Education*, 35: 242–248.
Hatano, G., Siegler, R. S., Richards, P. D., Inagaki, K., Stavy, R., and Wax, N. (1993). *The Development of Biological Knowledge: A Multi-national Study. Cognitive Development*, 8: 47–62.

Hayes, J. R. (1981). *The Complete Problem Solver*. Philadelphia: Franklin Institute Press.

Hein, G. E. (1998). *Learning in the Museum*. New York: Routledge.

Heit, E. (1994). Models of the effects of prior knowledge on category learning. *Journal of Experimental Psychology: Learning, Memory, and Cognition*, 20: 1264–282.

Heit, E. (1997). Knowledge and concept learning. In Lambert, K. and Shanks, D. (Eds.). *Knowledge, Concepts, and Categories*. Cambridge, MA: The MIT Press.

Helm, D. (1991). Neuro-linguistic programming: Gender and the learning modalities create inequalities in learning: A proposal to reestablish equality and promote new levels of achievement in education. *Journal of Instructional Psychology*, 18: 167–169.

Henry, F. M. (1953). Dynamic kinesthetic perception and adjustment. *Research Quarterly*, 24: 176–187.

Herschbach, D. R. (1995). Technology as knowledge: Implications for instruction. *Journal of Technology Education*, 7(1): 31–42.

Hestenes, D., Wells, M., and Gregg, S. (1992). Force concept inventory. *The Physics Teacher*, 30: 141–154.

Hewson, P. W. and Hewson, M. G. A' B. (1984). The role of conceptual conflict in conceptual change and the design of instruction. *Instructional Science*, 13: 1–13.

Higgs, J. and Titchen A. (1995). Propositional, professional and personal knowledge in clinical reasoning. In Higgs, J. and Jones, M. (Eds), *Clinical Reasoning in the Health Professions*. Oxford: Butterworth-Heinemann Ltd, pp. 129–146.

Hmelo, C. E., Holton, D. L., and Kolodner, J. L. (2000). Designing learning about complex systems. *The Journal of the Learning Science*, 9: 247–298.

Hofstein, A. and Rosenfeld, S. (1996). Bridging the gap between formal and informal science learning. *Studies in Science Education*, 28: 87–112.

Holton, G. (1975). Science, science teaching, and rationality. In Hook, S., Kurz, P., and Todorovich, M. (Eds). *The Philosophy of the Curriculum*. Buffalo, NY: Prometheus Books, pp. 101–108.

Holyoak, K. J. (1995). Problem solving. In Smith, E. E. and Osheron, D. N. (Eds), *Thinking an Investigation to Cognitive Science*, 3. 2nd ed. Cambridge, MA: The MIT Press.

Howard G. (1999). *The Disciplined Mind What All Students Should Understand*. New York: Simon & Schuster.

Howes, E. V. (2002). Learning to teach science for all in the elementary grades: What do perservie teachers bring? *Journal of Research in Science Teaching*. 39(9): 845–869

Hu, W. and Adey, P. (2002). A scientific creativity test for secondary school students. *International Journal of Science Education*, 24: 389–403.

Hunkins, F. P. (1989). *Teaching Thinking through Effective Questioning*. Norwood, MA: Christopher-Gordon.

Hurd, P. D. (1982). Scientific enlightenment for an age of science. *The Science Teacher,* 37: 13–15.

Huxley, T. H. (1893). On the educational value of the natural history sciences. In Huxley, T. H. *Collected Essays, III*, London, pp. 38–65. (repr. New York: Greenwood Press, 1968).

Inhelder, B. and Piaget, J. (1958). *The Growth of Logical Thinking from Childhood to Adolescence*. Trans A. Parsons and S. Milgram. New York: Basic Books (Original work published 1955).

Jarvis, T. and Pell, A. (2002). The effect of the challenger experience on elementary children's attitudes to science. *Journal of Research in Science Teaching*, 39: 979–1000.

Jarvis, T. and Pell, A. (2005). Factors influencing elementary school children's attitudes toward science before, during and after a visit to the UK National Space Center. *Journal of Research in Science Teaching*, 42(1): 53–83.

Jones, L. S. (1997). Opening doors with informal science: Exposure and access for our underserved students. *Science Education*, 81: 663–677.

Jones, G. M., Rua, M. J., and Carter, G. (1998). Science teachers' conceptual growth within Vygotsky's zone of proximal development. *Journal of Research in Science Teaching*, 35: 967–985.

Kali, Y., Orion, N., and Elon, B. (2003). The effect of knowledge integration activities on students' perception of the earth's crust as a cyclic system. *Journal of Research in Science Teaching*, 40: 545–565.

Kanari, Z. and Millar, R. (2004). Reasoning from data: How students collect and interpret data in science investigations. *Journal of Research in Science Teaching*, 41: 748–769.

Kelly, A. and Lesh, D. (Eds) (1999). *Handbook of Research Design in Mathematics and Science Education*. Mahwah, NJ: Lawrence Erlbaum.

Keys, C. W. (1994). The development of scientific reasoning skills in conjunction with collaborative writing assignments: An Interpretive study of six ninth-grade students. *Journal of Research in Science Teaching*, 31, 1003–10022.

Kisiel, J. (2005). Understanding elementary teacher motivations for science fieldtrips. *Science Education*, 1–20.

Klahr, D. (2000). *Exploring Science: The Cognition and Development of Discovery Processes*. Cambridge: MA: MIT Press.

Klahr, D., Fay, A., and Dunbar, K. (1993). Heuristics for scientific experimentation: A developmental study. *Cognitive Psychology*, 25: 111–146.

Koballa, T. R. and Crawley, F. E. (1985). The influence of attitude on science teaching and learning. *School Science and Mathematics*, 85: 222–232.

Kolodner, J. (1993). *Case-Based Reasoning*. San Mateo, CA: Morgan-Kaufmann.

Kolodner, J. L. and Leake, D. B. (1996). A tutorial introduction to case-based reasoning. In Leake, D. B. (Ed.), *Case-Based Reasoning: Experiences, Lessons, & Future Directions*. Menlo Park, CA/Cambridge, MA: AAAI Press/MIT Press, pp. 31–65.

Kosslyn, S. M. (1994). *Image and the Brain: The Resolution of the Imagery Debate*. Cambridge, MA: MIT Press.

Kubota, C. and Olstad, R. (1991). Effects of novelty-reducing preparation on exploratory behavior and cognitive learning in a science museum. *Journal of Research in Science Teaching*, 28: 225–234.

Kuhn, D. and Pearsall, S. (2000). Developmental origins of scientific thinking. *Journal of Cognition and Development*, 1: 113–129.

Kuhn, D., and Phelps, E. (1982). The development of problem-solving strategies. In Reese, H. (Ed.), *Advances in Child Development and Behavior*, Vol. 17. New York: Academic Press, pp. 1–44.

Kuhn, D., Amsel, E., and O'Loughhlin, M. (1988). *The Development of Scientific Thinking Skills*. Orlando, FL: Academic Press.

Kuhn, D., Black, J., Keselman, A., and Kaplan, D. (2000). The development of cognitive skills to support inquiry learning. *Cognition and Instruction*, 18: 495–523.

Kuhn D., Gracia-Milla, M., Zohar, A., and Anderson, C. (1995). Strategies of knowledge acquisition. *Monographs of the Society for Research in Child Development*, Serial No. 245, 60 (40), 1–28.

Kuhn, D., Schauble, L., and Gracia-Milla, M. (1992). Cross domain development of scientific reasoning. *Cognition and Instruction*, 9: 285–327.

Lam-Kan, K. S. (1985). *The contributions of enrichment activities towards science interest and science achievement*. Unpublished master's thesis, National University of Singapore.

Landies, D. (1980). The creation of knowledge and technique: Today's task and yesterday's experience. *Deadalis*, 109 (1): 11–120.

Langley, D., Ronen, M., and Eylon, B.-S. (1997). Light propagation and visual patterns: Preinstruction learners' conceptions. *Journal of Research in Science Teaching*, 34: 300–424.

Lave, J. and Wenger, E. (1991). *Situated Learning Legitimate Peripheral Participation*. Cambridge, MA: Cambridge University Press.

Lavoie, D. R. (1995). Preface. In Lavoie, D. R. *Towards a Cognitive-Science Perspective for Scientific Problem Solving*. The National Association for Research in Science Teaching, (NARST) Manattan, Kansas: Kansas State University. pp. iv-viii.

Layton, D. (1973). *Science for the People*. London: George Allen & Unwin Ltd.

Layton, D. (1994). *Innovations in Science and Technology Education*. Paris: UNESCO.

Layton, E. (1974). Technology as knowledge. *Technology and Culture*, 15(1): 31–41.

Leake, D. B. (1996). CBR in context: The present and future. In. Leake, D.B. (Ed.), *Case-Based Reasoning: Experiences, Lessons, & Future Directions*. Menlo Park, CA/Cambridge, MA: AAAI Press/MIT Press, pp. 3–30.

Lee, C. (1992). Literacy, cultural diversity, and instruction. *Education and Urban Society*, 24: 279–291.

Lee, C. D. (1999). Supporting the development of interpretive communities through metacognitive instructional conversations in culturally diverse classrooms. Paper presented at the annual meeting of the American Educational Research Association, Montreal, Quebec, Canada.

Lee, H. S. and Songer, N. B. (2003). Making authentic science accessible to students. *International Journal of Science Education*, 25: 923–948.

Linn, M. and Burbules, N. (1993). Construction of knowledge and group learning. In Tobin, K. (Ed.), *The Practice of Constructivism in Science Education*. Washington, DC: AAAS.

Lucas, A. M. (1983). Scientific literacy and informal learning. *Studies in Science Education*, 10: 1–36.

Lucas, K. B. (2000). One teacher's agenda for a class visit to an interactive science center. *Science Education*, 84: 524–544.

Luchins, A. S. and Luchins, E. H. (1970). *Wertheimer's Seminars Revisited: Problem Solving and Thinking* (Vol. 1). Albany, NY: State University of New York.

Maarschalk, J. (1988). Scientific literacy and informal science teaching. *Journal of Research in Science Teaching*, 25(2): 135–146.

Markman, E. M. and Callanan, M. A. (1983). An analysis of hierarchical classification. In Sternberg, R. (Ed.), *Advances in the Psychology of Human Intelligence* (Vol. 2), Hillsdale, New Jersey: Erlbaum, pp. 325–365.

Mayer, R. E. (1989). Models for understanding. *Review of Educational Research*, 59: 43–64.

Mayer, R. E., Dyck, J. L., and Cook, L. K. (1984). Techniques that help readers build mental models from scientific text: Definitions pretraining and signaling. *Journal of Educational Psychology*, 76: 1089–1105.

Mazur, E. (1992). Qualitative versus quantitative thinking: Are we teaching the right thing? *Optics and Photonics News*, 3: 38–40.

McCloskey, M., Caramazza, A., and Green, B. (1980). Curvilinear motion in the absence of external forces: Naïve beliefs about the motion of objects. *Science*, 210: 1139–1141.

McCloskey, M. (1983) Intuitive physics. *Scientific American*, 248(4): 122–13.

McDonald (1983). Technology beyond machines. In MacDonald S., (Ed.). *The Trouble with Technology*. London: Frances Pinter.

McDuffie, T. E. Jr. (2001). Scientists — geeks and nerds? *Science and Children*, 38: 16–19.

Medrich, E. A., Roizen, J., Rubin, V., and Buckley, S. (1982). *The Serious Business of Growing Up*. Berkeley: University of California Press.

Metz, K. E. (1995). Reassessment of developmental constraints on children's science instruction. *Review of Educational Research*, 65: 93–127.

Metz, K. E. (1998). Science inquiry within reach of young children. In Fraser, B.J. and Tobin, K.G. (Eds), *International Handbook of Science Education*. Part one, pp. 81–96.

Miller, G. E., Abrahamson, S., Cohen, I. S., Graser, H. P., Harnack, R. S., and Land, A. (1961). *Teaching and Learning in Medical School*. Cambridge, MA: Harvard University Press.

Minsky, M. (1985). *The Society of Minds*. New York: Simon & Schuster.

MIT Committee on Engineering Design. (1961). Report on engineering design. *Journal of Engineering Education*, 51: 647.

Mitcham, C. (1994). *Thinking Through Technology. The Path Between Engineering and Philosophy*. Chicago: The University of Chicago Press.

Monaghan, J. M. and Clement, J. (1999). Use of a computer simulation to develop mental simulations for understanding relative motion concepts. *International Journal of Science Education*, 21: 921–944.

Monaghan, J. M. and Clement, J. (2000). Algorithms, visualization and mental models: High school students' interaction with a relative motion simulation. *Journal of Science Education and Technology*, 9: 311–325.

Moore, G. E. (1903). *Principia Ethica*. London: Cambridge University Press.

Moscovici, J. (1998). *Shifting from activity mania to inquiry science — What do we (science educators) need to do*? Paper presented at the Annual International Conference of the Association for the Education of Teachers in Science/Minneapolis, MN, January, 1998.

Mostow, J. (1983). Machine transformation of advice into a heuristic search procedure. In. Michalski, R. S., Carbonell, J. G., and Mitchell, T. M. (Eds), *Machine Learning: An Artificial Intelligence Approach*. Vol. 1. Los Altos, CA: Morgan Kaufmann.

Mulholland, J., and Wallace, J. (1996). Breaking the cycle: Preparing elementary teachers to teach science. *Journal of Elementary Science Education*, 8(1), 17–38.

Murphy, G. L. and Lassaline, M. E. (1997). Hierarchical structure in concepts and the basic level of categorization. In Lambert, K. and Shanks, D. *Knowledge, Concepts, and Categories*. Cambridge, MA: The MIT Press.

Musgrove, F. and Batcock, A. (1969). Aspects of swing from science. *British Educational Psychology*, 39, 320–325.

Nash, J. M. (1997). Fertile minds. *Time*, 3: 49–56.

National Research Council (NRC). (1996). *National Science Education Standards*. Washington, DC: National Academy Press.

National Science Board Commission on Precollege Education in Mathematics, Science and Technology. (1983). *Educating Americans for the 21st Century: A Plan of Action for Improving Mathematics, Science, and Technology Education for all Americans Elementary and Secondary Students so that Their Achievement is the Best in the World by 1995*. Washington, D.C.: National Science Foundation.

National Science Standards. The science as inquiry standards http://www.nap.edu/readingroom/books/nses/html/6a.html#sis

Nemecek, S. and Yam, P. (1997). Science versus antiscience? *Scientific American*, 276, 96–101.

Newell. A. and Simon, H. A. (1972). *Human Problem Solving*. Englewood Cliffs, NJ: Prentice-Hall. (North Holland Publishers)

Novak, J. D. (1977). *A Theory of Education*. Ithaca, NY: Cornell University Press.

Novak, J. D. (1985). Concept mapping as an educational tool. *New Horizons for Learning's On The Beam*, 5(2): 4–5.

Novak, J. D. (1990). Concept mapping: A useful tool for science education. *Journal of Research in Science Teaching*, 27: 937–949.

Novak, J. D. and D. Musonda. (1991). A twelve-year longitudinal study of science concept learning. *American Educational Research Journal*, 28(1), 117–153.

NSES (1996). Available at http://www.nap.edu/readingroom/books/nses/html/6a.html#sis.

Orion, N. (1993). A model for the development and implementation of field trips as an integral part of the science curriculum. *School Science and Mathematics*, 93(6): 325–331.

Orion, N. and Hofstein, A. (1994). Factors that influence learning during a scientific field trip in a natural environment. *Journal of Research in Science Teaching*, 31(10): 1097–1119.

Orr, H. A. (1999). An evolutionary dead end? *Science*, 285, 343–344.

Paivio, A. (1986). *Mental Representations: A Dual Coding Approach*. Oxford: Oxford University Press.

Park, R. L. (2000). Magical thinking. Issues in Science and Technology Online. http://www.nap.edu/issues/16.3/br_park.htm

Parker, J. and Spink, E. (1997). Becoming science teachers: An evaluation of the initial stages of elementary teacher training. *Assessment & Evaluation in Higher Education*, 22: 17–31.

Parkyn, M. (1993). Scientific imaging. *Museums Journal*, 93(10): 29–34.

Patton, M. Q. (1990). *Qualitative Evaluation and Research Methods*, 2nd ed., Newbury Park, CA: Sage.

Pedretti, E. (2002). T. Kuhn meets T. Rex: Critical conversations and new directions in science centres and science museums. *Studies in Science Education*, 37: 1–42.

Perkins, D. (1992). *Smart Schools: From Training Memories to Educating Minds*. New York: Free Press.

Piaget, J. (1926). *La représentation du monde chez lienfant*. Paris: Alcan.

Piaget, J. (1946). *Les notions de movement et de vitesse chez l'enfant*. Paris: Presses Universities de France.

Piaget, J. (1969). *The Child's Conception of Time* (A.J. Pomerans, trans.) New York: Ballantine Books. (Original work published 1927).

Piaget, J. (1987). Possibility and necessity: Vol. 2. *The Role of Necessity in Cognitive Development* (H. Feider, trans.). Minneapolis: University of Minnesota Press. (Original work published 1983).

Piaget, J. and Inhelder, B. (1975). *The Origin of the Idea of Chance in Children* (L. Leake, P. Burrell, and H. D. Fishbein, trans.). London: Routledge & Kegan Paul. (Original work published 1941).

Piaget, J., Inhelder, B., and Szeminska, A. (1952). *The Child Conception of Number* (trans. C. Gattengo and F. M. Hodgson) London: Routledge & Kegan Paul. (Original work published 1941).

Pinker, S. (1997). *How the Mind Works*. New York: Norton.

Polanyi, M. (1962). *Personal Knowledge: Towards a Post Critical Philosophy*. Chicago: University of Chicago Press.

Pope, R. (2005) *Creativity: Theory, History, Practice*. New York: Routledge.

Popper, K. (1959). *The Logic of Scientific Discovery*. New York: Harper and Row.

Popper, K. (1963). *Conjectures and Refutations*. London: Routledge and Kegan Paul.

Prifster, H. and Laws, P. (1995). Kinesthesia 1: Apparatus to experience 1-D motion. *Physics Teacher*, 33: 214–220.

Raffini, J. P. (1993). *Winners Without Losers: Structures and Strategies for Increasing Student Motivation to Learn*. Upper Saddle River, NJ: Prentice Hall.

Rahm, J. (2004). Multiple modes of meaning-making in a science center, *Science Education*, 88(2): 223–247.

Ramey-Gassert, L. (1996). Same place, different experiences: Exploring the influence of gender students' science museum experiences. *International Journal of Science Education*, 18, 903–912.

Ramey-Gassert, L., Walberg. H. J., III, and Walberg, H. J. (1994). Re-examining connections: Museums as science learning environments. *Science Education*, 78: 345–363.

Reiner, M. and Gilbert, J. (2000) Epistemological resources for thought experimentation in science learning. *International Journal of Science Education*, 22: 489–506.

Reiner, M. and Gilbert, J. K. (2004). The symbiotic roles of empirical experimentation and thought experimentation in the learning of physics. *International Journal of Science Education*. 26(15): 1819–1834.

Rennie, L. J. (1994). Measuring affective outcomes from a visit to a science education centre. *Research in Science Education*, 24: 261–269.

Rennie, L. J. and McClafferty, T. P. (1996). Science centres and science learning. *Studies in Science Education*, 27: 53–98.

Rennie, L. J., and Williams, G. F. (2002). Science centers and scientific literacy: Promoting a relationship with science. *Science Education*, 86: 706–726.

Rennie, L. J., Feher, E., Dierking, L. D., and Falk, J. H. (2003). Toward an agenda for advancing research on science learning in out-of-school settings. *Journal of Research in Science Teaching*, 40(2): 112–120.

Resnick, L. B. (1987). Learning in school and out. *Educational Researcher*, 16: 13–20.

Rico, G. (1983). *Writing in the natural way: Using right-brain techniques to release your expressive powers*. Los Angeles: J.P. Tarcher.

Riggs, I. M. and Enochs, L. G (1990). Toward the development of an elementary teacher's science teaching efficacy belief instrument. *Science Education*, 74: 625–637.

Riley, D. and Kahle, J. B. (1995). *Exploring students' constructed perceptions of science through visiting particular exhibits at a science museum*. Paper presented at the annual meeting of the National Association for Research in Science Teaching, San Francisco, CA.

Rissland, E. L., and Skalak, D. B. (1991). CABARET: rule interpretation in a hybrid architecture. *International Journal of Man–Machine Studies*, 34: 839–887.

Rix, C. and McSorley, J. (1999). An investigation into the role that school-based interactive science centres may play in the education of primary-aged children. *International Journal of Science Education*, 21(6): 577–593

Roberts, P. (1994). The place of design in technology education. In Layton, D. (Ed.), *Innovations in Science and Technology Education*, Vol. V, pp. 171–179.

Roth, W. M. (1995). Affordances of computers in teacher-student interactions: The case of Interactive Physics™. *Journal of Research in Science Teaching*, 32: 329–347.

Roth, W. M. (2001). Learning science through technological design. *Journal of Research in Science Teaching*, 38: 768–790.

Ruffman, T., Perner, J., Olson, D. R., and Doherty, M. (1993). Reflecting on scientific thinking: Children's understanding of the hypothesis-evidence relation. *Child Development*, 64: 1617–1636.

Russel, R. (1995). What is informal science teaching? *Informal Science Review*, 14(1): 8.

Ryan, R. M. and Deci, E. L. (2000). Intrinsic and extrinsic motivations: classic definitions and new directions. *Contemporary Educational Psychology*, 25: 54–67.

Ryle, G. (1949). *The Concept of Mind*. London: Hutchinson.

Sahlins, M. (1976). *Culture and Practical Reasoning*. Chicago: Chicago University Press.

Sandifer, C. (2003). Technological novelty and open-endedness: Two characteristics of interactive exhibits that contribute to the holding of visitor attention in a science museum. *Journal of Research in Science Teaching*, 40(2): 121–137.

Sanger, M. J. and Greenbowe, T. J. (1997). Common student misconceptions in electrochemistry: Galvanic, electrolytic, and concentration cells. *Journal of Research in Science Teaching*, 34: 377–398.

Schank, R. C. (1996). Goal-based scenarios: Cased-based reasoning meets learning by doing. In Leake, D. B. (Ed.), *Case-Based Reasoning. Experiences, Lessons & Future Directions*. Cambridge, MA: MIT Press.

Schauble, L. (1990). Belief revision in children: the role of prior knowledge and strategies for generating evidence. *Journal of Experimental Child Psychology*, 49: 31–57.

Schauble, L. (1996). The development of scientific reasoning in knowledge-rich contexts. *Developmental Psychology*, 32: 102–119.

Schauble, L., Glaser, R., Duschl, R. A., Schulze, S. and John, J. (1995). Students' understanding of the objectives and procedures of experimentation in the science classroom. *Journal of the Learning Sciences*, 4: 131–166.

Schauble, L., Klopfer, L. E., and Raghavan, K. (1991). Students' transition from an engineering model to a science model of experimentation. *Journal of Research in Science Teaching*, 28: 859–882.

Schoeneberger, M. and Russell, T. (1986). Elementary science as a little added frill: A report of two case studies. *Science Education*, 70: 519–538.

Schon, D. (1987). *Educating the Reflecting Practitioner: Toward a New Design for Teaching and Learning the Professions*. San Francisco: Jossey-Bass.

Schwab, J. J. and Brandwein, P. (1966). *The Teaching of Science*. Cambridge, MA: Harvard University Press.

Segal, S. J. (1971). Processing of the stimulus in imagery and perception. In S. J. Segal (Ed.), *Imagery: Current Cognitive Approaches*. New York: Academic Press, pp. 73–100.

Senge, P. M. (1990). *The Fifth Discipline: The Art and Practice of the Learning Organization*. New York: Doubleday.

Sfard, A. (2000). Steering (dis)course between metaphor and rigor: Using focal analysis to investigate the emergence of mathematical objects. *Journal for Research in Mathematics Education*, 31(3), 296–327.

Shepard, R. N. and Cooper, L. A. (1982). *Mental Images and Their Transformations*. Cambridge MA: MIT Press.

Shepard, R. N. and Feng, C. (1972). A chronometric study of mental paper folding. *Cognitive Psychology*, 3: 228–243.

Shepard, R.N. and Metzler, J. (1971). Mental rotation of three-dimentional objects. *Science*, 171: 701–703.

Shore, R. (1997). *Rethinking the Brain: New Insights into Early Development*. New York: Families and Work Institute.

Shortland, M. (1987). No business like show business. *Nature*, 328: 213–214.

Shrigley, R. L. (1974). The attitude of preservice elementary teachers toward science. *School Science and Mathematics*, 74: 243–250.

Shulman, L. E. Knowledge and teaching: Foundations of the new reform. *Harvard Educational Review*, 57: 1–22.

Skamp, K. and Mueller, A. (2001). Student teachers' conceptions about effective elementary science teaching: A longitudinal study. *International Journal of Science Education* 23: 331–351.

Smith, J. P., III, diSessa, A. A., and Roschelle, J. (1993). Misconceptions reconceived: A constructivist analysis of knowledge in transition. *Journal of the Learning Sciences*, 3: 115–163.

Sodian, B., Zaitchik, D., and Carey, S. (1991). Young children's differentiation of hypothetical beliefs from evidence. *Child Development*, 62: 753–766.

Solomon, J. (1993). Reception and rejection of science knowledge: Choice, style and home culture. *Public Understanding of Science*, 2: 111–120.

Solomon, J. (1994). Towards a notion of home culture: Science education in the home. *British Education Research Journal*, 20: 565–577.

Solomon, J. (2003). Home-school learning of science: The culture of homes, pupils' difficult border crossing. *Journal of Research in Science Teaching*, 40(2): 219–223.

Sorensen, R. (1992). *Though Experiments*. New. York: Oxford University Press.

Spooner, W. E. and Simpson, R. D. (1979).The influence of a five- day teacher workshop on attitudes of elementary school teachers toward science and science teaching. *School Science and Mathematics*.79, 415–420.

Starkes, J. and Allard, F. (Eds) (1993). *Cognitive Issues in Motor Expertise*. The Netherlands: North Holland Publishers.

Stavy, R. (1991). Using analogy to overcome misconceptions about conservation of matter. *Journal of Research in Science Teaching*, 28: 305–313.

Stepans, J. and McCormick, A. (1985). A study of scientific conceptions and attitudes toward science of prospective elementary teachers: A research report. (ERIC Document Reproduction Service No. ED 266024).

Sternberg, R. J. and Lubart, T. I. (1999). The concept of creativity: Prospects and paradigms. In Sternberg, R. J. (Ed.). *Handbook of Creativity*. Cambridge: Cambridge University Press.

Stevens, C. A., and Wenner, G. (1996). Elementary preservice teachers' knowledge and beliefs regarding science and mathematics. *School Science and Mathematics*, 96: 2–9.

Stevenson, J. (1994). Getting to grips. *Museums Journal*, 94(5): 30–31.

Stone C. A. (1998). The metaphor of Scaffolding: It's utility for the field of learning disabilities. 31(4), 344–364.

Storksdieck, M. (2001). Differences in teachers' and students' museum field-trip experiences. *Visitor Studies Today*, 4(1), 8–12.

Strauss, A. L. (1987). *Qualitative Analysis for Social Scientist*. New York: Cambridge University Press.

Strauss, S. (1998). Cognitive development and science education: Towards a middle level. In Damon, W. (Series Ed.), Sigel, I. E. and Renninger, K. A. (Vol. Eds.), *Handbook of Child Psychology: Volume 4, Child Psychology* (5th ed). New York: Wiley, pp. 357–399.

Superior Committee on Science, Mathematics and Technology Education in Israel. (1992). *Tomorrow 98*. Publications Department, Ministry of Education, Culture and Sport, Jerusalem Israel.

Tamir, P. (1990). Factors associated with the relationship between formal, informal, and nonformal science learning. *Journal of Environmental Education*, 2(2): 34–42.

Thomas, N. J. T., (1999). Are Theories of Imagery Theories of Imagination? An Active Perception Approach to Conscious Mental Content. *Cognitive Science*, 23: 207–245.

Tobin, K. and Tippins, D. J. (1993). Constructivism as a referent for teaching and learning. In Tobin, K. (Ed.), *The Practice of Constructivism in Science Education*. Hillsdale, NJ: Lawrence Erlbaum & Associates. Chapter 1, 3–21.

Tosun, T. (2000). The beliefs of preservice elementary teachers towards science and science teaching. *School Science and Mathematics* 100: 374–379.

Toulmin, S. (1960). *The Philosophy of Science: An Introduction*. New York: Harper & Brothers.

Tschirgi, J. E. (1980). Sensible reasoning: A hypothesis about hypotheses. Child development, 51: 1–10.

Tunnicliffe, S. D. (1997). School visits to zoos and museums: A missed educational opportunity? *International Journal of Science Education*, 19(9), 1039–1056.

Tunnicliffe, S. D. (2000). Conversations of family and primary school groups at robotic dinosaur exhibits in a museum: What do they talk about? *International Journal of Science Education*, 22(7): 739–754.

Ulam, S. (1976) *Adventures of a Mathematician*. New York: Charles Scribner's Sons.

Viennot, L. (1979). Spontaneous reasoning in elementary dynamics. *European Journal of Science Education*, 1(2): 205–221.

von Glaserfeld, E. (1993). Questions and answers about radical constructivism. In K. Tobin (Ed.), *The Practice of Constructivism in Science Education*. Hillsdale, NJ: Lawrence Erlbaum, pp. 22–38.

Vygotsky, L. (1978). *Mind in Society; The Development of Higher Psychological Processes*. Cambridge, MA: Harvard University Press.

Vygotsky, L. S. (1933/1978). The role of play in development. In Cole, M., John-Steiner, V., Schribner, S., and Souberman, E. (Eds), *Mind in Society*. Cambridge, MA.: Harvard University Press.

Vygotsky, L. S. (1934/1986). *Thought and Language*. Translated and edited by Alex Kozulin Cambridge, MA.: MIT Press.

Wallace, J. and Louden, W. (1992). Science teaching and teachers knowledge prospects for reform of elementary classrooms. *Science Education* 76: 507–521.

Weinburgh, M. (1995). Gender differences in student attitudes towards science: A meta-analysis of the literature from 1970–1991. *Journal of research in Science Teaching*, 32, 387–398.

Wheatley, G. H. (1991). Constructivist perspectives on mathematics and science learning. *Science Education*, 75: 9–21.

Wheatley G. H. (1995). Problem solving from a constructivist perspective. In Lavoie, D. R. (Ed.). *Towards a Cognitive-Science Perspective for Scientific Problem Solving*. The National Association for Research in Science Teaching, Manattan, Kansas: Kansas State University, pp. 1–12.

Wills, P. (1977). *Learning to Labor: How Working Class Lads Get Working Class Jobs*. New York: Columbia University Press.

Windschitl, M. (2002). Framing constructivism in practice as the negotiation of dilemmas: An analysis of the conceptual, pedagogical, cultural, and political challenges facing teachers. *Review of Educational Research*, 72, 2: 131–175.

Wolins, I. S., Jensen, N., and Ulzheimer, R. (1992). Children's memories of museum field trips: A qualitative study. *Journal of Museum Education*, 17: 17–27.

Wolpert, L. (1992). *The Unnatural Nature of Science*. London: Faber and Faber.

Wolpert, L. (1997). In praise of science. In R. Levinson and J. Thomas (Eds), *Science Today*, London: Routledge, pp. 1–21.

Wood, D., Bruner, J. S. and Ross, G. (1976). The role of tutoring in problem solving. *Journal of Child Psychology and Psychiatry,* 17(2): 89–100.

Wymer, P. (1991). Never mind the Science, feel the experience. *New Scientist*, October: 49.

Yates, G. C. R. and Chandler, M. (2001). Where have all the skeptics gone? Patterns of new age beliefs and anti-scientific attitudes in preservice elementary teachers. *Research in Science Education*, 30: 377–387.

Zimmerman, C. (2000). The development of scientific reasoning skills. *Developmental Review*, 20: 99–149.

Zohar, A. (1999). Teachers' metacognitive knowledge and the instruction of higher order thinking, *Teaching and Teacher Education*, 413–429.

Zohar, A. and Nemet, F. (2002). Fostering students' knowledge and argumentation skills through dilemmas in human genetics. *Journal of Research in Science*, 39: 35–62.

Zuzovsky, R. and Tamir, P. (1989). Home and school contributions to science achievement in elementary schools in Israel. *Journal of Research in Science Teaching*, 26(8): 703–714.

AUTHOR INDEX

SUBJECT INDEX

Note to the reader: Page numbers appearing in italic refer to illustrations.